PHYSICAL LAWS
OF THE MATHEMATICAL
UNIVERSE:
WHO ARE WE?

PHYSICAL LAWS OF THE MATHEMATICAL UNIVERSE: WHO ARE WE?

NEETI SINHA

ARCHWAY
PUBLISHING

Archway Publishing books may be ordered through booksellers or by contacting:

Archway Publishing
1663 Liberty Drive
Bloomington, IN 47403
www.archwaypublishing.com
1 (888) 242-5904

Because of the dynamic nature of the Internet, any web addresses or
links contained in this book may have changed since publication and
may no longer be valid. The views expressed in this work are solely those
of the author and do not necessarily reflect the views of the publisher,
and the publisher hereby disclaims any responsibility for them.

Any people depicted in stock imagery provided by Thinkstock are models,
and such images are being used for illustrative purposes only.
Certain stock imagery © Thinkstock.

ISBN: 978-1-4808-2048-7 (sc)
ISBN: 978-1-4808-2049-4 (hc)
ISBN: 978-1-4808-2050-0 (e)

Library of Congress Control Number: 2015948842

Print information available on the last page.

Archway Publishing rev. date: 11/20/2015

☷ CONTENTS

INTRODUCTION

The conversation presented here is an offshoot of a stubborn yet pleasurable pursuit. The maneuvering through which this communication has materialized at the most fundamental level has served me with one of the most yearned-for occupations—connecting the "hidden" of the subliminal plane to the "obvious" of the observed universe; delineating the structure of an all-encompassing space-time, in which the cosmic arena, the quantum world, and the "aware" elements all reside agreeably.

The course of my scientific interests first led me to the scopes of biochemical and bio-molecular disciplines, and then deep into the theoretical fields of biophysics. In between, interspersed, were quite a few mini transitions that involved handling projects in which the two successive scientific curiosities didn't seem to even remotely connect. And after some such abrupt switches, I stumbled on a subject that is not only exceedingly methodical but also the most revelatory of the nature of reality that many of us crave to realize. From the position of this subject, the colors of different academic areas that I have glanced seem to converge to cast an unbroken spectrum of a learning order. The expositions from this revelatory field apparently authenticate the nature of an all-inclusive reality most deftly and with sharpest accuracy; they are the insights that emanate from the school of physical sciences.

The judgments of physical sciences, which I am trying to grasp, can lead seekers to an enlightening intersection, the interpretation of which can potentially unlock the utmost truth of the universe. This intersection, which melds the voices of physics and mathematics with the tone of perceptivity, by and of itself composes an entire subject, utilizing verdicts and abstractions that radiate from physics and mathematics to discern the nature of an all-embracing reality.

The interest in systematically translating the nature of an ultimate order has in part to do with the scientific inclination of my bio chart, and in some measure has to do with what was instinctive—a sort of knotty uneasiness that spurred me to grope for clarity concerning the schematic of the outmost continuum. Many of us zestfully ponder meaning and purpose, and the quest can cast its shadow in the arena of science or under the umbrella of any other scholastic domes. My avocation to align the scientifically observed with that which is cognitively grasped transpired following my own realization of the latitude that sweeps as watcher, extended beside the texture of space and time—the instance commonly referred to as self-realization.

The move to channel musings into a manuscript came with the reflection that there isn't a better way to throw light on the diagram that has drawn my interest for so many years than to sculpt it into a methodical write-up. So when I shipped out my elapsing affection for protein science and embarked on an unknown journey into the realms of the mysterious, assaying the perspectives toward piecing the floating bits into a meaningful tonality, it was only to help and enliven my own self. Channeling those assays into the form of a book was yet another ball game. The cycles of rumination, scribbling, and composing eventually led to a draft that perhaps could fit standards of a revelatory communication.

The current form emerged after rounds of construction, demolition, and restructuring, and involved major overhauls, with a great deal of chiseling, honing, tuning, and accentuating. Most of the insights that became fundamental in systematizing this communication, however, surfaced during the process of revisions, apparently providing a sharper and deeper understanding into the preliminary topic. The theme is not just juicy but also holds the key to unlock the greatest mysteries of the reality. The subject matter isn't just about how the universe is structured; it is also about how the elements of perceptivity sweep through the décor of the fully adorned space-time, where the consciousness too is pieced onto the design.

As I stepped along with the intent of peering into the utmost nature by the postures of the scientific purviews, my focus inadvertently fell on the kinesics of the subject of mathematics. At first I was purely amazed at how parallel mathematics is to the architecture of the universe that we see. Then, the more I probed into this numerical field, the sharper the image of the universe it printed. My attempts to converge the mathematical tone with averments of physics and elements of mind sprung open from then on. It is in the transparency of the suggestive force that mathematics carries that I bumped into a graphic by which the immersion of aware elements into the unified field of space-time can be shown.

In delicate ways we are all aware that, whereas the statements in physics broadcast the thoroughgoing view of the universe, there is no easy way to infuse consciousness within the same elaborative scheme. So, although the novel inspections do act as magnifying glasses to peek into the highly resolved what's what, to eyeball the absolute landscape, we need a rather suggestive window casement—one that hammers out when traces of mathematics, physics, and sentience are merged into a coordinated, well-ordered singleton.

Here I have tried to present a graphic of a full-length space-time that precipitates with the studs of the subliminal elements evenly peppered over the contour of the reality—the reality that we grasp from scientific observations. As I move forward into the subject matter, I shall try to furnish the script with apropos scientific findings and inklings, and with my own inferences about them, toward advocating the schematic of the absolute makeup.

Thus, the texture of the rendering presented in this book mushrooms from the pouring forth of three different springs: the mathematical tonalities that throw light on the fabrication of the reality, the appearance of space-time the way ratiocinations in physics portend, and the interjection of the preceding into an outline that accounts for the way we sense the universe. The resulting sketch must not just explain the scientific observations; it should also match the absolute complexion that is perceived cognitively.

I am hoping that the description that I present here not only exposes the gazing window into the ultimate but also shows how the inflection of the absolute space-time can be upheld by the frequencies of physics and the modulations of mathematics.

MARGINALIA

The review presented here attempts to tailor interpretations from widespread meadows of cosmological and quantum physics into a concerted tapestry. While venturing into the intermediary peripheries of the experimentally seen and the inwardly witnessed, I stumbled—many times, actually—onto the scientific peculiarities that, although fairly indicatory, just by the window of empirical logic appear gravely enigmatic.

Following are a few of these peculiarities:

- Why doesn't the image that the surveillance of the cosmic arena brings forth fully overlap with the drawing that the inspection of the quantum realm releases? After all, both the planes expose the temper of the very same universe. And so why do the peculiarities that turn up in quantum probing, such as the show of antimatter and the plaster of symmetry that arrive by the play of matter–antimatter camaraderie, stay ensconced in the spread of the celestial plane?

- How do we paraphrase the idea of gravitational singularity that appears in studious arguments as a mark in space-time where the forces of gravitation are infinitely sharp and the flow of time stays suspended?

- Why does the rhythm that is captured at the infinitesimal extent of the quantum realm tend to transmit the reality of the universe with further accuracy compared to what the macrocosm directly parades?

In this communication, along with attempting to delineate the ultimate trueness that encompasses space-time, the way R and D uncovers, the roots of the subconscious and the engaged consciousness, I shall also attempt unweaving a few apparent puzzlements, including those listed above, which loom in empirical judgments, by the practice of the very same proposition—reconciling mathematics, physics, and the way we perceive.

CHAPTER 1

Reflections of the Universe

Reverberations from the Domains of Physics,
Mathematics, and Perception: A Prelude

Perplexing Queries

The longing to venture into the concatenated frontiers of a rather circumferential topic hails from my quest to clarify the same set of perplexities that imprints virtually every human mind.

What is the essence behind the wondrous opulence that presents itself as the amaranthine scope of the endless universe? How did the physical laws that bring about the totality of this universe originate? In what manner do we hang about as a part of the universe's all-inclusive structural coordination? How do we explain the subtle mysteries and patterns we identify in our own lives? From what sources do our inner longings and volitions emanate?

The principles of physics by which the universe operates are the same ones through which we, too, live and function; we are the part of the same framework. Even more so, the paraphernalia of mind, an indispensable part of the conscious abode, also in certain inconspicuous ways circuits the same fabric. Hence, how do we decipher the mechanics of communion between the design of the universal structure and the spread of the intellective miscellany, within the range of an all-exclusive circuitry?

Scientific certifications and the theorems that revolve around them proffer rich chunks of well-defined illuminations, but only in an assortment of disconnected answers, each of which belongs to its respective specialized field of research—where two genres of messages often do not sprucely converge. And thus, the facts from different experimental setups do not congruously join in offering a tuneful rationale behind the show of the full-blown ball game—an operation that clasps all the elements into a single unified field of concordance. There exist bafflements that plague scientific discernments as much as they do the bounds of philosophical appetites toward inferring life and universe. These bafflements may begin to disentangle only if a proper diagram can be summarized in which the *conscious* and the tagging *subliminal* habitants prevail within the order of the macrocosmic totality, in a methodic fashion the way physical sciences broadcast.

In the braid of the space-time continuum, each quantitative entity—including the singletons that wander in the latitude of the cognizance—rolls about both as space and as time. Accompanying the course of a conscious jaunt swings the faculty to perceive. By the endowment of discernibility we get the hang of the fact that we as embodiments also turn up, in some implied way and to a certain degree, beside the definitiveness of space-time; that is, beyond the fringes of the physical extent. Apart from just throbbing as elements of space-time, a conscious life at a fine-spun level also simultaneously senses the passage of time. (One should, more properly, refer here to space-time. This will become clear shortly, after we assimilate Einstein's space-time theories.) We sense space; we sense time. We might at some level exist beyond both. The traces of elapsed space-time, in the field of life, project in forms of vivid memories and nostalgic imprints. Thus, at a very basic level, the pack of conscious force is channeled by the confluence of three ingredients: space, time, and the sweep of the perceiver. One might also say that pure witnessing occurs seemingly independently of the other two ingredients: the space and the time.

In this presentation we strive to see a methodic order in which the space-time continuum soaks up the role of our senses as well. In doing so, we may arrive at a picture in which the role of consciousness axiomatically emerges from the scientific landscape of reality, further offering a means for reconciling certain discrepancies that we currently encounter in interpreting reality.

One of the topmost essential elements of reality is relentless change. Constant change is the most common principle regardless of where and at what scale we look—cosmic, quantum, subliminal, and within each of us. All is constantly flowing and reconfiguring. Let's start our journey with a little glance at the aspect of this indispensible transition.

The Enigmatic Permanence of Transitions in the Scope of Reality

Whether picturing the imprint of a living thing or the dint of sheer matter, the judder of reality tags along behind the drumming of imperishable changes and limitless shifts. The 1990 launching of the Hubble Space Telescope led to the confirmation that not only is the sphere of the cosmos ballooning with prodigious velocities, but also that the expansion itself is accompanied with the allotments of quickening. The rate of hastening is determined to be about fifty to one hundred kilometers per second per megaparsec. (One megaparsec is equivalent to the distance of approximately 3.2 million light-years). This is known as the Hubble constant. It connotes that a celestial object that is one megaparsec distant travels away from the observers with the speed of fifty to one hundred kilometers per second, while a cosmic body ten megaparsecs away would recede ten times more rapidly. Likewise, quantum-scale examinations firmly declare that the substratum of materiality shifts and remodels against every scintilla of realism. The permanent state of transitions also is uniformly mirrored in the fields of the supraliminal and the subliminal. From stellar objects to quantum particles to day-by-day settings to a smorgasbord of phrenic and emotive verities, every little thing abides in a constant state of flux and transfiguration. The only never-failing homogeneity in the theatricals of existence is the permanency in the propagation of change itself.

One might probe the programming behind the state of constant flux in the arena of pure physicality, or relentless shifts in the sphere of consciousness, after subsuming the attributes of the subconscious into the same expanse of realism—the one that strikes the eye. Including the subconscious is essential, for we know that the structure of physicality is determined by the manifestation of the living receptivity, with which comes the blobs of subconscious, sweeping the circuitry of the same space-time continuum.

The display of the concrete that meets the eye must accompany the spray of the dispersed. The well-known interpretation in systematic studies, phrased as the wave–particle duality, suggests that the clank of particle and the smear of wave are the two separate quirks of the very same delivery. Having dawned in the seventeenth century, this quantum mechanical view, secured over the works of Christiaan Huygens and Isaac Newton, asserts that along with every concrete embodiment lugs the shadow of the dispersed. The continuum of definite and abstract is an inevitability evoked behind the occupancy of every entity—including those that coat the arena of awareness. This continuum also is manifest in the mold of the methodic coliseum that the macrocosm displays, in its entirety. An overarching definite–abstract twoness must be contrived to picture an all-encompassing universal dynamic.

The Foundational Queries

The idea of apprehending the precipitation of intelligent life by the description that research and inquiries present is not altogether a novel one. Investigators at the vanguard of methodical understanding heedfully aim to understand the structure of ultimate reality—that which soaks the truth of consciousness as well. In most of such undertakings, however, the functionalities of the cerebral, which is utilized to discern factuality, are not accounted at any level, let alone at the most fundamental one. Aside from the framework of scientific insights, the philosophical windows employed to make out the utmost reality, conversely, do not tend to absorb the regard of realism in a most stringent manner, the way exact sciences advertise. Fusing the two can show how our own placements, too, pop into reality fixed within the forces of electromagnetism and gravity. The segregation of the two—experimental observations and cognizance—leads to distinctive scientific dubieties and mazy mathematical puzzlements, not to mention the blockage of the route that could extend into an ultimate view. Such an ultimate view not only would aid in ironing out the blunt scientific conundrums, but would also administer a full-blown sketch of the all-encompassing landscape that we all ardently long to fathom.

Following are some of the dilemmas that plague current methodical deductions, along with a few commonplace ponderings:

- the extension of infinite steps toward cracking Zeno's paradox, the quandary of classification in Russell's paradox, and the conceptualization of infinity
- the monumental mystification that mathematical assertions meticulously parallel the exactitude of the physical world
- the superfluous dimensions, instead of the regular three, that ooze in a current rationalization toward discerning reality
- the show of antimatter in quantum mechanical makeup, while in the vastness of the cosmos the shadows of antimatter are nowhere to be found

- the appearance of the gravitational singularity—an infinitely wrapped spatial event—where the throb of time lay abeyant
- the unequivocal inhabitance of variegated parallel universes in quantum demarcations, when cosmic landscape appears to be steadfastly singular
- the spelling out of the duality of the particle and its comrade wave

Here are some of the most basic questions one may ask:

- How do we come to be?
- Why is the universe the way it is?
- Why, in the twirls of solidity, does the conception of God diffuse?
- How can the progression in the biological belt be sewn into the inflation of the voluminous universe?

Finally, we may ponder this:

- How can we piece the constituents of mind into the continuum of the universe that we notice and learn?

Some of these plights at this point may appear to readers as slightly specialized. The reason for citing them here is to highlight the delicate and obvious bewilderments that we come across, not just toward limning the current of the consciousness in the inferable flexions of space-time but also in coming up with the ultimate design by way of experimental scrutiny. All of the nubs listed above join in different ways to disentangle some of the nethermost secrets— subjects upon which we will come, subsequently, in specific contexts.

Only by taking account of the skirts of sentient presence enlaced along the adages of the fundamental principles can science transparently enter into pageants in which we can optimistically embark upon the task of untangling the bedazzlements that cloud scientific eruditions—whether over the nature of quantum mechanics and cosmological physics, toward contriving the decoding of mathematical conundrums, in locating our own placement in the massive latitude of the universe, or against decrypting the apperception of God.

I myself have been part of specific research enquiries that serviced varied areas of biophysical research. Each of these fulfilling projects in a particular way has furnished a rung in the learning ladder that gravitates toward gauging the fibers of the truth. Envisioning the absolute workings of the universe in totality, however, would involve blending the research inputs with the portraitures of perceptivity and meaning.

The lusty thirst for the conclusive knowledge may begin to gratify if we can think up an equation that embraces the receptivity within the same macrocosmic framework that the exact disciplines endorse.

In the rest of this chapter, I try to throw in some of the most critical bits and pieces that are essential in constructing a holistic picture—which we will endeavor to build from the next chapter on. Thus you may find that the subjects of the successive subheadings lack contiguity. The point here is to bring together the most crucial features from three different sectors: physics, mathematics, and the perceptive nature.

Procuring a Panoptic Window:
The Picture from Physics

The physical principles of the natural world that we pick up from scientific research, radiating from the branches of classical and contemporary physics, occupy every nook and corner of the universe squarely—be it a speck of a sub atomic cast, the shadow of the entire universe, the playing of the parallel universes, or the nature of space-time that permeates the umbrella of consciousness. It is because the coming about of the physical forces is synchronous to the launching of matter, and the principles that describe the streaming of matter in a corner of the quantum realm are the same as those that impart the pondering of matter in remote nooks of gigantic galaxies, or in the mode by which our own frames are positioned in the curvature of space-time—all orchestrating in concordance in a unanimous systematic makeup. An audible example to point up the homogeneity in universal behavior is the manner of dynamism that is showcased by the behavior of matter; the mass entities persist by the means of constant duress along a center, both at the atomic and the cosmic levels (figure 1.1).

From stark illuminations by the studies of classical mechanics—that describe the ordinance of dynamism at the macroscopic scales, including the astronomical bodies, instituted by Isaac Newton—ascended the magnum opus of relativity, drawing in the most effortless as well as one of the most revelatory idioms, $E = mc^2$, conveyed by the celebrated physicist Albert Einstein in the beginning of the twentieth century. Beside the most worthy in-depth insights that this equation emits—some of which we will see later—at the most modest level it utters that mass and energy are two names for the very same game. And although the idea of relativity was debuted in the formulation $E = mc^2$, establishing that the beat of time and promptitude of movement exist in relativity, the phenomenon of relativity evidently permeates the absoluteness in myriad divergent ways, including when we take account of the subliminal accompaniments in the design of the whole—the argument we will later run into—between the universal renditions and mental portrayals as well. The same formulation was furthered to incorporate the gravitational force, again by Einstein, delivering the understanding that space-time conforms to the force of gravity; that which appears to be a vast magnitude of flatland is contoured under the influence of the basic forces (figure 1.2).

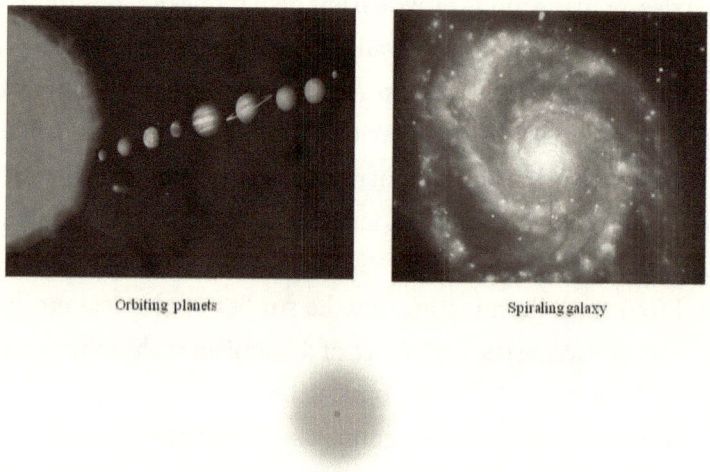

Orbiting planets   Spiraling galaxy

An atom: A simplistic depiction of electron whizzing about the nuclear center

Figure 1.1. Impetus about the center.

[Images i and ii are downloaded from GRIN (Great Images In NASA)]

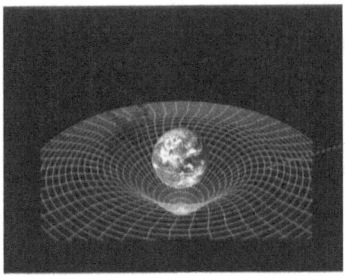
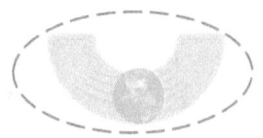
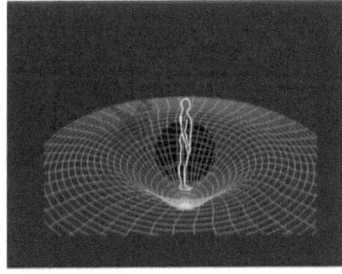

Figure 1.2. Space-time is curved by the anchor of every concrete form.
Space-time depiction credit: NASA; Earth image credit: NASA's Earth Observatory/NOAA/DOD

With the onset of quantum mechanics in the beginning of the twentieth century, instituted primarily around the works of Max Planck, Albert Einstein, Niels Bohr, Louis de Broglie, Satyendra N. Bose, Wolfgang Pauli, Werner Heisenberg, Paul A.M. Dirac, Enrico Fermi, Erwin Schrödinger, and Richard Feynman, spanning the time period from about 1900 to around 1928, and its significant subsidiaries, quantum field theory and, recently, the archetype of string theory—the figuration of all the formerly vested interpretations into a single paradigm—myriad previously unknown mysterious traits of reality made the scene.

For instance, the contemporary findings exposed that the bed of space-time is exhibited with ceaseless undulations, that the spatial extent is exhaustively thorough (that is, there isn't any empty space present between mass–energy embodiments), that there nest the casings of parallel universes within the shell of the concluding makeup, that every bit of matter hangs around with the parcel of an antimatter, that the existence swings with the mark of duality and the imprints of symmetry and supersymmetry, that the baffling instances of rips and tears make their way within the sheet of space-time, that the universal fabric assimilates the enigmatic vista of black hole and the thrilling invisible notional avenue of wormhole. These apparently mystical quirks in the accents of the universe in fact by and large implement a way toward capturing a complete arena—and we will apply them over and over in striving to peek into the integrated chime of the absolute makeup.

Fully seeped within the tapestry of the grasped quantum realm is a peculiarity that suggests that the accuracy of outcome is concomitant to the act of observation. A spectacular occurrence defined as wave function collapse, the formulation of the Copenhagen interpretation put forward by Niels Bohr, hints that a wave that is structured by all possible eigenstates—plainly an eigenstate refers to a specific position, out of all possible, in the landscape of the wave—evolves to acquire one of the eigenstates upon confronting an observer. The observer and what is observed are two facets of the very same coin—an etching of relativity between the beholder and what is beheld, within the solidity of the exhaustively networked terrain of reality.

It is on the foundations of relativity and quantum principles that the modern physics grows and refines, and our understanding of the universe emerges and improves.

In consideration of avoiding the percolation of too many technical details, especially in the headmost section, for now it will suffice to bring in just one more detail: that the mass–energy, in all varied forms, exists in concordance with the fundamental forces—at the most rudimentary level, the electromagnetic and gravitational. For instance, at any given point in time, a cosmic body is positioned in space-time according to the universal potentialities that radiate around it. The entities in the radius of the perceptive arena too are subject to the natural potentialities; the perceiving of the world is made possible only because the amiable arrangement of electro-magnetism and gravity burrows us into a specific befitting seat in the all-seats-consumed coliseum of the networked reality (figure 1.2; figure 1.3).

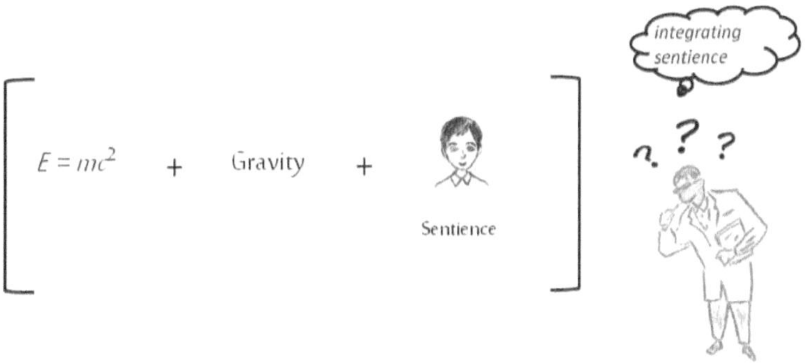

$$\left[\; E = mc^2 \;+\; \text{Gravity} \;+\; \underset{\text{Sentience}}{\text{🙂}} \;\right] \quad ?\,?\,? \quad \text{(integrating sentience)}$$

Figure 1.3

It is in the behavior of the quantum bounds that we can see how the "sentient presence" infiltrates the reach of the same continuum, and carries on by the forces of the same over arching universal principles. Whether attempting to rationalize the unveiling of the parallel universes, the occupancy of antimatter, or the emblems of duality and symmetry in ambits of colossal cosmos, the elements of the subliminal belts, along with the indispensable stretch of *consciousness*, automatically fall onto the stage of the whole.

The fusion of universal and mental, and thus our own position-ing in the network, can be further appreciated in the window of the

observer. It is the sheer purity of the observer that is outright realized, radiating as an independent field alongside the reticulation of
space-time, cognitively by the instance of self-realization. And it is
important to note that the fundamental conceptions of physics too
necessitate the presence of an observer in their descriptions.

The waving of the parallel universes, the steadiness in the prodigious inflation of cosmic scene, and the fresh idea of the signature
of dark energy that is conjectured to obliterate gravitation cannot
be justified, again, in the absence of the window into which the
observer sights. And I am optimistic that we will subsequently
acknowledge, via the depiction that is rolled out here, that the reference of the observer in physical theories corresponds to the realm
of the observer that is appreciated psychically.

And when we interfuse the elaborate tapestry of space-time
that the sciences declare with the tenancy of the pure observer we
will notice that not only do the disparities between quantum and
cosmic judgments start to vanish, but also that the spectrum of the
animate attendance makes its appearance within the panorama of
the same space-time vista.

Furthermore, in the display of such an integrative painting, we
may be able to find the rationale behind the grand mystification
concerning why the speech of mathematics mirrors the order of
truth—and thus, alongside, the basis of why certain mathematical
conundrums come into play can also be appreciated.

The concept that the three-dimensional universe de facto is a
holographic projection of an ambit that itself is two-dimensional—
currently flaring in contemporary views of the unified theory[1]—
can be seen to be just in an all-encompassed painting that accommodates the subliminal and wears the span of consciousness, where
the scene of space-time rolls in through the establishment of two

distinct windows: one at the electromagnetic level that the eye interviews, and the second from the vantage point of the observer who witnesses the very same design of space-time, manifested in a form of a rather linear sweep—which has burrowing within it the additional spatial extents. Whether the geometry of three-dimensional scope pitches sharper image or the countenance of the two-dimensional extension alongside the observer casts higher explicitness would depend on the location of the window, but the crucial underlining fact is that the pack of information that is utilized in the actualization of any form of façade—displaying the labyrinthine circuitry of space-time—is one and the same.

The Premade Visages of Mathematics

In all likelihood it appears that as opposed to mathematics being an invented technique against the demand to codify, parameterize, and reason—the phrasings of mathematics sort of breeze in within the valleys of the reality for themselves to be discovered. I use the term *discovered* because everything that we lay eyes on throughout the breadth of nature and far-reaching horizons of the universe reverberates emblems of mathematical design in myriad ways, to such a great degree that it is as if the masquerade of the full-length universe is composed in the clear-cut language of mathematics. And it does not stop there. Our psychological parcels (for example, artistic cognizance and aptitude, the visceral sense of geometrical proportions) also loom in mathematical jottings. The numerical aphorism also pervades the constitutions of physique and structuring that underlie biological composition. The stage setting of the entire physicality makes a show of numerical spectacle, and the carriages of mind seem to break through, side by side, within the same numerical framework.

History substantiates that the keen involuntary penchant for mathematical inquiries is often coupled with intuitive zest to see the hidden truths by the speech that the numeral complexion broadcast. Plato (ca 400 B. C.), Gottfried W. Leibniz (1646–1716), René Descartes (1596–1650), Georg F. Cantor (1845–1981), Bertrand Russell (1872–1970), Kurt Gödel (1906–78), Henri Poincaré (1854–1912), and Alfred Tarski (1901–83) are only a few of the renowned names, out of a long list of personalities, who ventured to bridge the philosophical inclination of truth to the materiality of mathematics—not to mention those anonymous ones in primeval times who followed mathematics for its acute depth in cluing into the nature of the actuality.

The concept of zero, which conceivably first appeared in Indian mathematics—the arithmetical flavor first seen by Indian mathematicians Brahmagupta, in the seventh century, and Mahavira and Bhaskara, in the fifth and twelfth centuries; the symbolized ∞, the appellation originating from the Latin word "*infinitas*" (signifying limitless extent), the allusive front of which has been reckoned since ancient Indian and Greek times and, much later, in the late nineteenth century, the concept of which was formulated in numerical terms by German mathematician Georg Cantor, leading to the inception of set theory; the most basic of arithmetical tempers; the banners of exponentiations and fractionations; the concealed pith in the matrix of the set theory; the spatial expression by the description of trigonometry; the flexions of plane and advanced geometries; the intricacies of exceedingly complex algebraic behaviors—which are magnificent tools in understanding recondite physical phenomena; all, in one way or another, whisper the hidden facets of the actual universe and how it executes.

Alongside the above-listed carvings, however, there remains

one of the greatest amazement—the ubiquitous impressions of number patterns in all aspects of nature. The matching of mathematical articulation with the way the universe behaves is far from just being a conceptual notion. Some of the most authentic behaviors of space-time in the arena of existence—such as the prime presence of antimatter, and the cast of symmetry and super-symmetry—were first discovered in mathematical formulations and then validated by the experimental procedures. The topological folds that come about by the sleek expression of complex algebraic equations help decipher the intricate theoretical purviews, such as the theories of relativity, the structure of hyperspace, the track of the manifold, and, most recently, the lens of the string theory.

Some recent titles, such as *Is God a Mathematician* and *The Golden Ratio*, by Mario Livio;[2,3] *The Mathematical Universe*, by William Dunham;[4] *Emblems of Mind*, by Edward Rothstein;[5] and *Why Beauty Is Truth*, by Ian Stewart,[6] extend to disseminate the premise of mathematical adumbration in the way of the universe. The comparableness of the factual vocalizations and numerical speech is not just authentic but powerfully sharp—and there seems to be no ambiguity about this.

The accent of mathematics seems to hold a masterly key to unlock the hidden dimensions of the real world. Although the foundational enquiry as to why the numerical language carries such an efficacy has remained suspended in oblivion, a subject that has perhaps been considered beyond the bounds of logic. The most awe-inspiring, as well as authentic, feature that the arithmetical tone utters is the hint that the currents of the psyche appear to be mathematical too, airing the same calibration that the universe arrays. It is as if the mental elements are inclined toward the gestures that the universe simultaneously exposes.

When we account the elements of mind as discrete numerical entities floating in a unified pool of systematic concordance, the expression of mathematics singlehandedly evolves to bring out the very same portraiture of space-time to which the most advanced formulations of physics point. Mathematics insinuating reality seemingly has to do with the impression of the observer that it brings along in each of its dicta; we will snap back to this theme over and over.

Melding the Psychological Realm with the Mural of the Espied Universe

The spurs of feeling and the imprints of memory that make the scene in the radius of the mental extent are vividly extant—albeit within the constitution of the cognizant. Anything that pitches a tent in any alcove of existence exposes a tincture of an entity and therefore must, in some way, narrate the occupancy of the mass–energy (or physical) aspect (figure 1.3). If the accompaniments of psyche can be envisioned to be systematically tied into the fabric of the materialistic plane that the eye espies, then a unified system that assimilates awareness can be contrived.

The ingredient of abstraction in the mix of existence is the fact that soaks the modern views of scientific discernments. *Abstract* signifies diffused, stretched, and wavelike continuity features. The concrete–abstruse accompaniment is core to the nature of space-time, and this coordinative attribute heightens the urge to integrate the tempers of awareness in the photograph of ultimate realness in identifying an all-integrated snapshot. This is purely because what manifests as discrete to the eyes shades the spread that materializes in the window of the pure observer—a spread that encompasses the mental attributes.

The idea of wave–particle duality, from the doctrines of Huygens and Newton; the notion of uncertainty in quantum mechanics, initially positioned by Werner Heisenberg; and the complementarity principle, which overlaps the Copenhagen interpretation and was deployed by Niels Bohr—addressing the distinctiveness of particle and wave, highlighting complementarity in the amalgamated duality of particle–wave—call for the accounting of the sweeping space-time spread in the window of the pure observer, and the assimilation of the accompanied subliminal bearings alongside the cosmic ones. In this unified view, the continuum of the universe and the inner plane—that of the observer—would coincide with the sequence of concrete cosmos and the whole-hog quantum. Cosmic and quantum aren't two different planes; just two different observing windows. I will provide a clearer picture on this later on.

To restate, the all-hemmed-in view that materializes from the vantage point of the observer concatenates the universal and subliminal make-ups, according the sweeping space-time continuum the way we learn from quantum studies. This may seem a little baffling at this point, but we will explore this with much rigor later on.

And it is in accounting the occupancy of the observer that the streaming of the parallel universes we grasp from empirical judgments too can be appreciated and supported. And it can't be emphasized enough that the scene of the pure observer can be reasoned only after we have accounted all the subliminal elements within the tapestry of space-time physicality. I will provide further elaboration on this in chapter 6.

A related deliberation that is signifying in this context is the shadow of duality—that one thing can be seen in two different ways—which arrives in many different ways in academic interpretations. Physical duality is a slick way to an overarching conformity of space-time, which assimilates the mental elements as well. The schematic of duality, however, at an apical level that accounts the observer, would reveal a scene of an utmost concordance—a state of nonduality, of the matter and its forces, that we shall come upon in concluding chapters.

A few of the above notations may sound a little specialized, especially for the context of this chapter. I, nonetheless, still wanted to point out the niches we will be voyaging about in our journey to sketch out the tracery of an utmost space-time.

The Residence of Conscious Realm in the Trajectory of the Space-Time

In 1916, over the theory of general relativity, Einstein pointed out that the light curves along the force of gravity. In the concrete plane, the face-to-face encounter between two objects is dictated by the sweep of universal forces between them—including the gravitational force. In other words, once we acknowledge that, in the plane of physicality, the entities exist only in relation to one another, the way in which the plane is cast—stretched or contoured—becomes irrelevant. In the intricate grid of space-time, we, in due course, will appreciate that the flow of the physical continuum echoes the breezing of the abstruse, which has elements of the subliminal stitched within the same universal tapestry. Under the sight of the pure observer the concrete and the abstract, as well as the physical and the mental (or, more accurately, the inner), go on, leagued, to dispatch a seemly superstructure of an all-hemmed-in stretch.

The layout of relativity in the cosmic regime is calibrated by gravity—the basic force discovered by Isaac Newton in the seventeenth century as a force enabled between all forms of matter that is directly proportional to the mass, and inversely proportional to the square of the distance between the mass entities—in the sense that in the arrangement of the celestial arena the positioning of each and every astronomical article is dictated by every other cosmic form, to an extent that is determined by the mass and the intervening distance. Overall, for revolving cosmic objects of similar proportions, the gravitational influence between the centered star and the one orbiting closer will be higher compared to the gravitational pull exerted between the stellar midpoint and the one revolving farther out. Thus, in the array of the universal display, matter abides by the grid of gravitational sinuosity, emanating via the presence of each one of the embodiments, which in turn regulates every other one, with varying intensity. In the vista of the observer such a sweep of space-time by design would ferry the subliminal elements, emotive and intuitive, as well.

It will help if we seize a mechanism that can straightforwardly indicate the forces by which the cognizant engage within the decree of the physical principles. Newton's third law of motion from classical mechanics suggests that forces come in action–reaction pairs; the two interacting bodies simultaneously experience enforcements that are equal in magnitude and opposite in direction. The precept holds equitably irrespective of where in space-time the seed of the paired extortions appear; that is, it includes the dealings in the perceptive ambits as well. The acts to propel open a door, drag about a box, draw up a painting, fling a Frisbee, and shake hands, all involve forces that come in two opposing pairs—for the purpose to be served. The instance of the door being pulled open is a consequence of the exertion of the duress on one side and the door counteracting that potential with the same magnitude on the other. Lifting the painting from the floor to the wall again involves

the potencies of two sources—an obvious one is employed to ele-
vate the painting upward, and the other one is the gravitational pull
that combats the action equitably. If one of the forces is missing,
the other cannot be exercised.

Deeply ingrained in this principle is the relentless ritual that
states that every act, as well as every behavior in the arena of the
universe, engages interactivity. The pitch of action–reaction concomitancy reflects the most stubborn form of symmetry that per-
colates the universe. Envision a nudge opening a gate, without the
counteractive resistance that radiates from the gate, or drawing
up a picture from the floor, without gravitation. Such a unidirectional force cannot be effectuated, plainly because the act of force
can be applied only against a counteractive potentiality for a deed
to become apparent in the palpable show of the universe. At the
reflective level, the concerted deal of action–reaction under the
psychological awning would then directly correspond to the spur
of feelings against every worldly enactment that we encounter. By
the same token, the elements of the emotive realm also participate
equitably in the same interweaving where the fibrils of world and
consciousness stay on par.

The subjective furnishings, such as creative pursuits and personal interests, hang about knitted in the absolute space, whereas
structures in accordance with the universal texture become manifest only under witnessing by the latitude of the observer. And in
the field of vision that proliferates alongside the plain observer, the
display of the physical and the bird's-eye view of the all-enveloped
sweep—which also assimilates the elements of the phrenic plane—
would gel as a cascade of one and the same flow.

Now let's gather some of the assorted pickings that are seemingly pivotal in piecing together a universal image. We will explore

the clear advancements made, and obvious traps, in physics toward capturing a unified scheme. We will use both advancements and traps to our advantage toward seeing a full-length picture of reality.

The Highly Resolved Image of Unified Field in Modern Physics, and the Accompanied Conflicts

Appending the revelatory mathematical concepts with suggestive scientific observations may roll out an engraving of totality that is not explicitly evident through just the microscope of a single discrete experiment, or by the glance of a solitary conception. Picturing all-in-all collectivity might reasonably be the essence toward understanding the ultimate nature.

Ever since the onset of conscientious endeavor that Einstein kindled in contriving a unified view—which he chose to call the unified field theory—to soak up electromagnetism, gravity, and all of the quantum mechanical behaviors that were deemed established at that time, it has become commonplace to strive to consolidate all of the clues into a single comme il faut picture. And despite the mounted information that comes from the zones of advanced quantum and astronomical cadres, attempting to throw together a unified scheme continues to be one of the major challenges of modern understanding.

But the present-day enterprise is channeled with additional sets of input that were not conjectured in Einstein's time. Although these sets provide a firmer foundation, they make the whole deal slightly compounded. The copious bulk of supplementary declarations that have made their appearance since Einstein's time mainly come from the areas of particle physics, quantum behaviors, and the avant-garde formulations of mathematical physics.

The inputs from the tracts of particle physics have mounted

with the crew of subatomic particles becoming evident, where even the proton and neutron of the atomic nucleus are shown to be made out of a variety of subparticles known as quarks. The swarm of additional particles, such as neutrino, muon, tau, and the particles denoting forces, the photon (electromagnetic), gluon (strong nuclear), weak gauge boson (weak nuclear), and the recent Higgs, and hordes of antiparticles are also found to be swimming in the field of certainty.

By about 1970, though the insight of the unified field had not been fully gained, all the preceding scoops hitherto were seen to blend in to parcel out a scheme commonly known as the standard model of particle physics, which assimilated the particulate nature with three of the four fundamental forces—leaving the gravitational at bay, as it did not seemly interlock.

The popular science titles *The Elegant Universe*, by Brian Greene,[7] and *The Grand Design*, by Stephen Hawking and Leonard Mlodinow,[8] confer crisp summaries on relevant phases that led to the understanding of the present-day enrichments of physics.

The present judgment is that the melding of the quantum-stationed quirks, fastened with the standard model of particle physics, and general relativity (the influence of gravity), would spit out a formulation that would broadcast the canvas of the unified field, in which every ingredient of the field complements and supplements the brilliance of an overarching phenomenon. An integrative equation that can sharply point to the theory of everything at the ultimate level would show that nature follows a single underlying mode, no matter how microscopic or telescopic the field at which we peer.

Long after Einstein's induction to gauge the theory of everything, three fundamental forces (the electromagnetic, weak, and strong nuclear forces) out of the four were seen to successfully conform to the quantum description of the microscopic particulate world, with the fourth force, gravity, left aside from the conceived carving of the factuality. This representation, as we earlier noted, was elevated to the designation of the standard model of particle physics. However, even today, there remain transparent disparities between the pictures of the universe that emerge from different branches of physics—especially the ones that exude from quantum and cosmic regimes.

We will turn back to how these differences cue the arrangement of the apical stage in the more elaborative chapters. For now it is sufficient to note that the diversified portraitures of the realism do not hint toward disparate natures of space-time; they only insinuate the variegated ways that we have used to weigh the one and the same actuality.

Whether through the microscope of the subatomic arena, the eye lens of the world, or the telescope of the cosmos, whatever is observed is the imagery of truth—not small, not big, just image. The notion of distance, or size, is a property reflected in relativity. When we look at the quantum realm we see the constituents in relativity, the same way relativity is seen in the cosmic arrangement. Quantum and cosmic are just two independent observing panes reflecting the behavior of the universe; the size difference between the two realms cannot be taken to be conclusive, especially when gauging the mechanics of the universe.

Mathematics enthusiasts would savor that in the details of the current mathematical models of reality arrived at via the mode of string theory, we can appreciate that the extent of size we deduce would depend on the measuring parameter.[7]

Theoretically, the certainty of electromagnetism in the staging of the quantum field deviates from the directness of gravity in laying out the cosmic terrain. At the ultimate level of comprehension, both the forces must gesture toward one potentiality in some deep way, similar to the way it is perceived for the electromagnetic and weak forces. Relatively recently (1979), via the mode of the calibrated deductions of quantum fields, Sheldon Glashow, at Harvard University; Abdus Salam (1926–96), then affiliated to Imperial Collage, London; and Steven Weinberg, at the University of Texas, pointed out that out of the four fundamental forces, the weak and the electromagnetic forces, at the most fundamental level, are the attributes of the very same potentiality, which, suggestively, is known as the electroweak field.

The infusion of both gravity and electromagnetism within a single diagrammatic frame even now is one of the gravest scientific challenges. The "demanding" itself may evocatively whisper the scenario of the topmost trueness, where the arrangement of space-time in one window spells out the charge potentialities, uttering the electromagnetic force, and in the other the mass potentiality, uttering the gravitation—the two reflections of one amalgam in whipping up the spatial extent. In any case, from the vantage point of the watcher or plain observer, in the relativistic panoramic setup of space-time, the two forces come to be part of the very same pillar on which relativity rests.

The point isn't whether the outputs emanating from diverse branches dish out the exact same blueprints; the bottom line is how the assorted range of outputs can be utilized to envision an overarching schematic in which the disparate findings show up corroboratively—and their discrepancies resoundingly appear just. It is hoped that we are taking off to embark on the route that leads to the scene of the ultimate setup, where each and every bit of information has its own nestling place, and where the inputs from one scientific area do not mismatch with the intellection from the other—where the grounds of classical, quantum and up-to-date physics, the underpinnings of mathematics and the substructure of biology, do not contradict each other but stay on equal footing.

Here I am optimistic to undertake a voyage that all along its way draws out an unabbreviated scenery, accentuating how the discernments from the disciplines of classical physics to the all-enveloping quantum theories of strings, including each and every empirical catch, backed up by their mathematical counterparts, could point toward the systematic entirety. Consequently, we would step into a setting that weaves the consciousness into the realness of the universe and into the hyperspatial parallel extensions into which the universe stretches.

Let us now begin our journey with the following rudimentary assets in hand.

The Elemental State of Concurrence and Relativeness

Whether we talk of action–reaction law or action–reaction that spurs in the mental domain, an action and a reaction coexist. They are relative; one exists because of the other. Our understanding suggests that the dealings of the brimming universe, whether in the window of the cosmic or in the optical of the quantum, cannot be deemed carried out without the inherency of co-potencies that interfuse at all levels—from course images to shrouded subtleties. The package of any component comes encased in the cloth of concomitancy—mass against energy, space against time, time relative to movement, matter relative to antimatter, enthalpy in correspondence with entropy, particle in communication with wave—and,

in the contemporary proceedings, the matter allied to the forces—
and so forth.

Whereas the tinge of synchronicity in the declaration of action–reaction, at a finespun level, is slightly distinguishing from
the stately flags of concerted aspects that I note here, the mode of
coevality, nonetheless, radiates through every facet of reality, from
whatever evaluative plane. The coevality, in turn, shines a suggestive symmetry, the luminosity of which is further accentuated in
the quantum scenes of matter, which absorbs antimatter, and in the
integrated plot of the universe, which sponges up parallel universes.
Taken all together, the architecture of actuality tethers the shadows
of simultaneity by every slant and at all standings.

The emotive paraphernalia too vibrate in pairings—compla-
cence being the other end of discontent and the sense of the buoy-
ancy against the sense of heaviness, and so on. The essentiality of
twoness in every notch of nature puts a finger on the state of co-ex-
istence that underlies the palpating surface of actuality. And it is on
the pillar of the vital concurrence that we can see the fortification
that incorporates the continuation of the parallel universes, and the
alongside presence of the observer—also in simultaneity.

A few years ago I attended a seminar lectured by a theoretical
physicist at the Johns Hopkins University. It was a somewhat a
general notional talk on the subject of time that explored to bridge
the theoretical conceptions with the observations of experimental physics. His illustrations on the whole featured interpretations
about the textural description of the spatiotemporal sheet, show-
casing that the chief stratum is structured with the couplings of
cause and effect—engendering the weaves of localities, where the
adhesive force of a cause-and-effect neighborhood weakens as one
wanders outward—in the vastness of the spatial extent. Beyond the

globs of the cause–effect localities, in the sheet of the universe, he explained, lies a continual stretch that he called "God's eye."

The appearance of a universal eye in the all-girdled present-ment or the need for an observer in the methodical readings en-treats the seizing of a mural where the entire gamut of evidence pitches in to signify distinctive flavors that emerge from the same underlying postulates.

From now on, in rest of the text, the observer that lies side by side with space-time is described with varied appellations, indica-tively, as "the empty space," "the watcher" (or "the watching"), "the witnesser," or "the dimensionless plane," depending on the context of the subject.

It is the span of the pure witnesser—separate from physical, subliminal and emotive elements—that is identified as the upmost denouement at the juncture of self-realization. And perhaps we would realize along the voyage here that even this dimensionless expanse exists only in the partnership with the physical texture. The observer doesn't carry through independent of space-time— the stamp of ultimate concurrence in the rhythm of the universe.

There Isn't Any Entity Hovering beyond the Ply of the Physical Reality

Thus the diagram of the grand universal coherency can be seen in the surpassing extent where the materialistic concurrence palpates only along a plane that although stretches matter-free, blankets the threads of matter synchronously—the plane of the observer that beholds the corporeal theatricals. And so, in the conclusive de-sign, the emblems of entities deploy only within the interlacing of the physical principles, well warped through the beat of their own presence, including the ones that texturize the mental makeup.

Thus the timeworn premise that the laws of the physical universe are seen to be driven by a grand entity becomes an incoherent plot. It is as if we are looking for a bigger law that would operate a set of smaller laws.

Here we may be able to benefit from one of the mathematical articulations to envision the picture of reality—the conceptual interlinkage that I have elaborated in a later chapter. However, I must leave a note of caution that the comparableness between the signatures of mathematics and the shades of the universe is far deeper and brawnier than the notional reciprocation I post here, for here the introductory level demands more conjectural quoting than the more direct ones. Steadfastly interblended in the modulations of algebra resides the powerful apothegm of the polynomial—a numeral expression that brought immense expediency in the progression of the multitudinous fields in exact sciences. From basic formulations of pure mathematics to complex intonations of applied mathematics to the understanding of the deep-seated phenomena of the physical world, the articulation of the polynomial is an indispensable tool, with encyclopedic applications, in channeling the statements of physical sciences.

The polynomial expression utilizes the unknown (or the variable), most commonly symbolized as x, y, z; the coefficient, a number fused to the unknown; and the constant, just a number on its own. To cook up a polynomial declaration, these ingredients join together by arithmetical functions of addition, subtraction, and multiplication. The forbidden activity in the recipe of the polynomial is the variable carrying a negative exponent or acting as a divisor. The expression $x^2 - 2x - 8$ is a polynomial expression made out of three terms: x^2, $2x$, and 8 where x represents the variable, 2 the coefficient, and 8 the constant. The graph of a polynomial always depicts a smooth continuum of a predictable disposition. The smooth continuum is effectuated because of two reasons. All of the participating terms homogenously melt to bring out an expression that fairs the terms integrally, and the fact that the downsizing of x by the presence of a negative exponentiation or the x being a divisor is forbidden. The tied-in smooth sailing of the constituent terms may throw light on how the interactive configuration of the continuum in the real world might be taking place, where the unbroken flow of space-time would allude to knitted succession in universal functionalities that are doled out by the interactive manifestations of participating elements, handing out an integrative design, which is smooth because all the entities are homogenously involved and remain in continuum because the elements are warped under the influence of the fundamental forces.

Indeed, a much deeper glance is deemed necessary to get the true idea on how the polynomial accentuates the scene of reality—and we will try to access that in the later chapters. As a matter of fact, polynomial statements can show how the intonations of perceptivity can be seen stitched in the loftiest tapestry; we shall eyeball this in the eventual sections. Here this is only to identify how the behavior of the ultra recondite universe can be assimilated over rather facile speeches that the numerical language verbalizes.

The craving to fathom the complexion of the ultimate remains deeply ingrained alongside our emotive sense. So often I rummaged through the information philosophically offered on the wisdoms of the ultimate reality, toward seeing contextual meanings behind the colors that effuse in scientific deportments. I later explain a little about how I came across the instance of realizing the ultimate observer. Here I just want to drop a hint on the catalytic progression that culminated in the transmission intimated here.

Following the juncture of perceiving the all-pervading dimensionless plane, I obsessively and ardently rummaged for discoveries, theories, and concepts that would not only indicate the existence of such a plane, parallel to the bounds of the physical reality, but would also suggest a unifying plot that subsumes the span of this enveloping observer. It is at this point I became totally intrigued by the unerring appeal that is cast in the embodiment of mathematics—from simple formats to the most complex configurations. Mathematics, in its exclusivity, shows both: it highlights the presence of the formless plane, which lounges tethered with the spatters of space-time, and it indicates how mass–energy fabrications, including those that burrow through the conscious plane, symphonize—all in an overarching scheme that accommodates the side-by-side whirls of the parallel universes. The declarations of physics fluently harmonize within the sculpture of the same symphony.

The true revelatory potentiality of mathematics becomes stark when the constituents of the mind are held to be the forms of mathematical entities, in all of those articulations that either symbolize or communicate physicality, such as the notations of infinity in set theory, the diffusion of sleekness in polynomials, the seizing of higher dimensions in strings, or the grasping of the relativistic universe. Perhaps once we acknowledge the role of the five senses through which we come upon the web of the interactive physical laws, we may be able to contemplate the foundational queries of the ultimate what and absolute why. And if we can throw the emotive paraphernalia into the whole ball game, we may be able to see the luminous transparency of mathematics in highlighting not just how the physicality maneuvers but also how the averments of mind, too, permeate the same landscape that we measure as the picturesque reality.

The Metaphysical Obvious

The citation of the presence of a sole witnesser beyond the rims of space-time that appears in theosophical and philosophical disseminations—and that which is explicitly indicated in some of the ontological ancient Indian scriptures Upanishads—is indicative of the state that is perceived through the domains of the psychical. Also referred to as self-realization (or enlightenment), perceiving through the fields of mind implies acknowledging that a level aside the physical tangents does pound; it is the level of the perceiver or witnesser that boundlessly fuses with the panorama of space-time. The intricate complexion of space-time that empirical knowledge avers and continues to stratify brings forth the subtleties that, when fused with the measurelessness of *witnessing*, spit out a systematic paradigm by which the physical facts can be better appreciated—the physical facts that also infuse the continuum of the consciousness.

Beyond just acknowledging the deep-seated fact that the state of the metaphysical witnesser intimately merges with the realm of the physical, it is notably acute to limn how the absolute mass–energy dynamics proceed in the panoptic ambience, if we are to climb upward on the scientific ladder of understanding.

CHAPTER 2

Mathematical Reality of the Universe under the Canopy of Physical Principles

Physical Truth in Numerical Order

The image of zero in the morphology of mathematics symbolizes the ambience that sweeps behind the array of numerical entities. The definitiveness of numbers (or the language of mathematics) can commence only in the atmosphere of the empty space that zero betokens.

The elemental message of number 1 is the shine of oneness—the indivisible entirety that aggregates to cement the other numerals. As we count 2, 3, 4, …, the indispensable whole of 1 is three-dimensionally projected 2, 3, 4, … times. However, the outset of countability coincides with the manifestation of *gap* that intercalates in between the arithmetical elements for their distinctness to become apparent. The ambience of the empty space on which the digits commence is the same background by means of which they become distinct.

Very Plain, Yet Paradoxical

Recapped by Aristotle—"in a race, the quickest runner can never overtake the slowest, since the pursuer must first reach the point whence the pursued started, so that the slower must always hold a lead", Zeno's (Zeno of Elea, 490-430 B.C.) famous paradox on the race between Achilles and the tortoise ventilates a plight that Achilles will never be able to defeat the tortoise, if only the tortoise gets to start off just a little ahead. Under that condition, the tortoise begins at racetrack position B, while Achilles begins from position A (figure 2.1). Let's say the distance between point A and B is one meter. For Achilles to lead the competition, he would need first to leap the distance of 1 meter, and to do that he will have to, at the outset, pass over the 0.5-meter mark, and before achieving that, he will have to cut across the 0.25-meter mark, but before accomplishing that, he will have to sprint through the distance of 0.125 meter, and so on. Achilles will never be able to make any advancement, let alone traverse A to B, because there innumerable counts are nested between any two points on the track (figure 2.1). There isn't any smallest distance probable that Achilles should cross to make a progression. Thus the tortoise will always be at an advantage. The mounting of paradox is inevitable, for via the act of counting there is no limit as to how deep or broad one can go in seeing the digital elements.

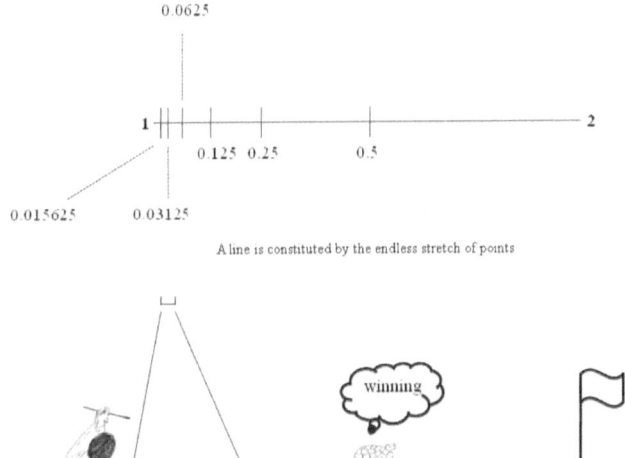

Figure 2.1 Achilles can never get ahead of tortoise because of the presence of infinite steps across any two points of the track

The quotas of the types of enumerations can be exacted; the numbers are systematically grouped, the way nested Russian dolls are structured, but as far as the appearance of a digital figure is concerned, there is no limit to how far away we journey looking for it. We will perpetually find one. The arithmetical entities are curled up through five stages, denoted as N, for natural numbers; as Z, for integers; as Q, for rational numbers; as R, for real numbers; and as C, for complex numbers. They remain sequentially nestled to embrace all the numeric planes. These five types of characters, as well as their order, are recollected by a popular, but not very stylish, mnemonic: "Nine Zulu queens ruled China." In the unfurling of the numeral dictionary, the N crops up first, and thus sits centrally, followed by Z, which curls up N and has additional features—the inclusion of 0 and negative numbers that are not available to the lot N. Z is further wrapped by Q, the lot that fell in when the poses of fractions made the scene and the rational numbers needed to be the part of the numeric pigmentation. The allotment of R burrows all the preceding and adds the tints of irrational numbers—the expressions that cannot be shown in clean-cut ratios.

The irrationals are of immense interest to the aim of this book, and I will follow them shortly. The outmost in the standard categorization is the area of C, which became part of mathematical texture when the square root of a negative number, or the multiplication of the two negative numbers, was required to produce the value in negative disposal only. Customarily the square root of both positive and negative numbers is a positive value, and the segregation of the positive from the negative became acute in the furtherance of mathematical evolution and in the understanding of the physical accuracies by the order of numbers, and we will see some of this later as well.

The mere feat of tallying brings about the launching of discrete forms. And for the pixel of the numeral draft to become distinct, the spread of the empty space in between digital boundaries must simultaneously appear.

The ideas of zero and infinity have incited artists, scientists, and philosophers. The concept of zero as an arithmetical denotation first appeared in ancient India around the seventh century A.D., although its philosophical linkages to the ultimate nature date back to the Vedic period that stretched around 1400 B.C., where the cognate idea of *shunya* (Sanskrit for "zero"), translating to "void" or "empty," was envisaged to state the reflection of metaphysical nothingness and emptiness, beyond the plane of the physical. Ordinarily, the idea of zero as nothing is indicative that number one, or any other number, for that matter, is bigger than zero. Basing the significance of empty space that zero incarnates in bringing about the mathematical phrasings, the conviction might be a little unfounded. Zero is not smaller than *one* or any number, no matter how large. The hum of zero betokens a presence that simultaneously comes to hold at the commencement of counting, mounted behind the shades of digitals. Impressionally, zero images the rendering of empty space that stretches behind the counts. It shadows a state of an unambiguous span that must aerate the ambience for the mathematical language to come alive.

The obsessive penchant of mathematicians to outline the temperament of infinity very likely has to do with the picture of truth that becomes apparent in the plane of numbers. There is indeed an unambiguous overlap between the unfolding of the digits and the physical entities. Both the advancements, of mathematics and the physical, entail a synced span of a formless actuality that continues fixed alongside the discrete traits.

This chapter is divided in two parts. The first section, "The mathematical allurement and the mathematician enchantment," touches on the general meaning, reality, and allurement of mathematics. In this context, we will also acknowledge some of the outstanding mathematicians who were incessantly drawn to contemplate the truthfulness in mathematical nature.

In the second section, "Mathematics Imaging Three-Dimensionality," we will begin to directly acknowledge the correspondence between numerical language and physical reality. The ways in which the arithmetical dispositions symbolize three-dimensionality and reveal hidden facts is the central gist of this section.

The Mathematical Allurement and the Mathematician's Enchantment

The Numerical Space

German mathematician Georg Cantor (1845–1918) is recognized as the founder of set theory, for his shrewd contributions in defining the number categories based on the spatial densities that different types of numbers carry. Possessed with a compelling desire to comprehend the nature of infinity and the mold of physical trueness, he suffered chronic depression much of the active period of his life, which included spells of hospitalization and sanatorium stays. During the bouts of despondency he would break off from mathematical work and bounce into philosophical ponderings to quell his agitation in gauging the essence of the apical reality.

His numerical competence highlighted that there exist different levels of infinities, depending on the spatial density a number set exhibits in relation to the other sets.

The numerical density or crowdedness has to do with the bulk of digits present within a given periphery. Whereas for integers, or whole numbers, the bulk of 1 through 5 indicates a record of five entities, for rational numbers (fractions) the load of entities within the same compass will be markedly higher, flourishing in forms of 1.2, 1.3, 1.4, 1.5, and so on (figure 2.2), and in many different sequential series. The unlimited supply of rational numbers will fill in the extent between any two integers; there are copiously more—in fact, as many as one wants—rational elements than integers. So, on the basis of the way the digit types present their spreads, the fractions are described as dense, as compared to whole numbers.

It is vital to mention that despite the higher density the rationals hold compared to the wholes, Cantor argued that the rationals and wholes still belong to the same infinite set. He did so by showing that one-to-one correspondence can be established between these two types of numbers. Both are countable.

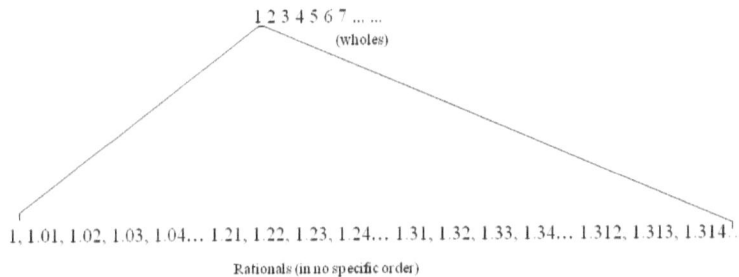

1, 1.01, 1.02, 1.03, 1.04... 1.21, 1.22, 1.23, 1.24... 1.31, 1.32, 1.33, 1.34... 1.312, 1.313, 1.314...

Rationals (in no specific order)

Figure 2.2. Rational numbers are denser than whole numbers.

The breadth of real numbers (R) would include further spatial extents because now the manifestation of crowdedness is exerted by integers, rationals, and irrationals. Irrationals stretch into transcendental extents; I will discuss them elaborately shortly.

The rational number line is denser because their superabundant cuts endlessly cram the numerical boundary. The disposition of the real number (R) amplitude is even more extravagant. The real numbers embrace irrational numbers as well, and so stand for an expanse that warps not only discrete perpetuity but the streaks of indefinite spread as well. Whereas a rational number simply represents a quotient or a fraction, such as 6.4 or 274/8, the irrational numbers announce the unlimited expansion of their decimal digits—never culminating or motioning the banners of numerical patterns in their continual proliferation. Thus the irrational numbers are all the more dense, as compared to the rationals or any other enclosed sets, for here even a single entity comes with ceaseless development of behavior. The square root of 2, $\sqrt{2}$, is the most commonly used example of an irrational number (expression 2.1). The others regularly cited are pi (π), the ratio of a circle's circumference to its diameter (expression 2.2); phi (ϕ), the golden ratio; and Euler's number, e, a mathematical constant commonly found as the base of the natural logarithm.

$$\sqrt{2} = 1.41421356237309504880 \dots \quad (2.1)$$
$$\&$$
$$\pi = 3.141592653589793 \dots \quad (2.2)$$

None of the preceding renderings can be etched by clear-cut brinks; the amplification never converges to furnish a discrete singleton to showcase exactness. The number 894/5 is a rational number because it results in an exact value, 178.8. 894/7 is also a rational number. It unrolls to be 127.714285,714285,714285 ..., prolonging with unending repeats of "714285." The appearance of the reruns categorizes this number as rational—a concrete behavior mirrored. The irrational numbers, conversely, do not parade a definitive structure; $\sqrt{2}$ belongs to the irrational allotment because it unrolls an unending furtherance that remains immune to any signature of periodicity. In a little while, we will be able to grasp the underlying cause responsible for such a phenomenon.

Because of the irresolute conduct in the unwinding of the irrational numbers, they are taken to be specks that are all the more jam-packed compared to the rational and the whole numbers; and in prescribing the infinite sets, Georg Cantor placed them in a higher order of occupancy. Thus, in the realm of definitiveness the irrational gamut echoes a nonconforming posture announcing the lack of an unambiguous framework—a recondite reach that permanently diffuses away from the resolved straight-out plane (figure 2.3). The guise of the mathematical irrationality is closely tied with the decorum of nature and the universe, from the quantum pulsations to cosmic regularity—which we will try to equate in the next chapter and the one after that.

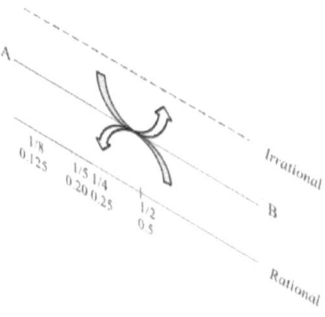

Figure 2.3. Irrational number space relative to rational number space. The two independent spatial realms within points A and B. Dotted irrational line represents inexactness, while whole and rational numbers reflect the settlement of discrete entities.

The Captivating Expression of Math

The sameness between physical architecture and numerical tones turns up in two ways: first by the graphical duplicates of the universal structure that numerical embodiments exhibit, and secondly the other way round—by the image of the mathematical dialect that we see in the throb of the universe.

While the trait of endlessness is reverberated over the tone of all number types, an integer epitomizes countable discrete object and an irrational banner uncountable dispersed appearance, and subsequently we shall see that the crests of irrationality in the substantiality of the universe are indicative of the prevailing relativity and concurrence, the way we catch on from the empirical judgments—the canopy of relativity, from Einstein's outlook, and the tones of concurrence, duality, and symmetry, spouting from quantum mechanical postulates. The ploys by the crystal-clear fringes of the whole numbers, on the other hand, by their very nature draw attention to the coming about of the peerless structure in the array of the universe—the way we eyeball. Thus the disparate genres of digital intonations broadcast diverse decisive traits of the realism—some outwardly obvious and others abysmally articulative.

But, before taking a plunge into the binary anatomy, let us take in a little more on the intellective bent of some of those mathematicians who wandered to enrich the scholastic fields, but also veered about inhaling the smell of the absolute in the mannerism of mathematics. Some, like Georg Cantor, as we have noted before, bore congenital mental challenges in their paths as well.

Around the context of set theory abides a name that is thought to be that of the most telling logician of the modern mathematical era. The name is Kurt Gödel (1906–78). Gödel was an Austrian American logician, mathematician, and philosopher. He made immense contributions in decrypting logic toward judging the foundation of mathematics, and his philosophical ken threw light on the descriptive nature of the mathematical entities, pointing to the notion of mathematical realism—that is, that the idioms of mathematics truly exist and are not just invented. The flavorsome reads on his philosophical inflections can be found online in the *Stanford Encyclopedia of Philosophy*. His analytical work—built on the framework that Georg Cantor laid—brought monumental advancements to the conspicuousness of the set theory.

Beneath the sharp wit lurked the afflicted clouds. In the posterior years of his life, Gödel too displayed signs of mental derangement. He caught himself in the clutch of obsessive thoughts that his food was being poisoned, and as a consequence, he chose to consume only the foods that his wife prepared. Once that source was no longer available, because of his wife's health issues, he chose to starve himself to death; he weighed only sixty-five pounds at the time of his death.

Georg Cantor was clasped with a deep-seated conviction about the sway of a higher power, and he went through jarred states of fanatical urge to disseminate his words on the conceptions of God and truth that he scented in the midst of his dissections of infinity. A piquant report on Cantor's hard-boiled devotion in excogitating the numerical fundamentals, and the nature of his out-of-the-ordinary individuality, is given in the title *The Mystery of the Aleph*, by A.D. Aczel.[1]

These are just two authorities who laid the foundations of hitherto untouched disciplines yet had bored, troubled mind states they couldn't transcend. Their compulsive analytical drives and the images they provided dazzle shades of beauty and truism that mathematical syntax whispers. The fanatical influence that the numerical articulations cast seemingly springs from the mystical shadow of truism, thus beauty, that mathematics reflects which in turn overcomes the emotive faculty of the ardent enthusiasts, much in the same way an artist remains captured by the allurement of creativity. And by the measures of mind, both math and art stand synonymous to beauty and truth.

Excerpted are some of their phrasings that echo the scholarly credentials and the associated thoughtful bents that these avid mathematicians carried:

Georg Cantor:

- "The essence of mathematics lies in its freedom."
- "In mathematics the art of proposing a question must be held of higher value than solving it."
- "I have never proceeded from any *Genus supremum* of the actual infinite. Quite the contrary, I have rigorously proved that there is absolutely no *Genus supremum* of the actual infinite. What surpasses all that is finite and transfinite is no *Genus*; it is the single, completely individual unity in which everything is included, which includes the Absolute, incomprehensible to the human understanding. This is the *Actus Purissimus*, which by many is called God."

Kurt Gödel:

- "To arrive at the totality of integers involves a jump. Overviewing it presupposes an infinite intuition. What is given is a psychological analysis. The point is whether it produces objective conviction."
- "The world is rational."
- "Concepts have an objective existence."
- "Intuition is not proof; it is the opposite of proof. We do not analyze intuition to see a proof but by intuition we see something without a proof."

The unending array of arithmetical entities brings forth an outline toward grasping the idea of infinity. However, in the cincture of physical reality, the delivery of infinity can be judged neither categorically nor without accounting the force of relativity and synchronicity. With the show of everlasting eternal universe comes the casing of the gaper—or, in the language of mathematics, the act of measuring. In allocating assemblages of infinities, Cantor proposed the notion of transfinite numbers—indicative of digits extending into the unknown, where the natural numbers are placed in the most primary transfinite number category, designated as aleph-null ($\aleph_0$). The next category of transfinite number is that of the real numbers (R), designated as aleph-one ($\aleph_1$). Both of these number categories mark infinities, yet the two sets are distinctive for their spatial attributes, as we have seen before. While the former signifies how the apparently infinite universe is presented in a definitive manner, the latter portends the mode of transcendence in the scaffold of actuality. The mode of transcendence is central to the transmission of space-time and is boldly reflected in the orders of classical and modern diagnostic reads. The astounding wonder of the transcendental numbers, as well as their recitation of factualism, is the core subject of the following two chapters.

The musings of zero and infinity under the umbrella of numerals seem to draw out two polar ends, where zero is considered smaller than one and infinity is the coming of boundlessness in the stretch of space-time. Apart from the arithmetical designations, however, while zero indicates the gap that is necessary in the presentment of the digital boundaries, the endlessness of infinity signals the continuum of "gauging," or, in the tonality of physics, the witnessing of the observer. Contemplating over the unabridged tenor of the supreme infinity, Georg Cantor came up with the inkling of absolute infinity that he tagged as the "ultimate comprehensiveness," which transcended even the transfinite numbers that he had utilized to group the levels of rather direct infinities. The quandary of defining infinity sets in as soon as we attempt to characterize limitlessness independent of the framework of relativity, and the casing of intercommunications that comes with it.

This, and some of the other topics covered in this chapter, may sound a little basic, especially for all those who are familiar with the theme of how mathematics exists and evolves. I still wanted to initiate the topic from the most fundamental step before we head into the relatively complex constitution in journeying to view the reciprocity between the speech of mathematics and the theme of the absolute makeup that links the customs of consciousness.

The notion that mathematical gestures image the anatomy of the universe is not brand-new. Labeled as the mathematical universe hypothesis, by Max Tegmark from the Massachusetts Institute of Technology, it addresses the mathematical structure of physical reality. Tegmark has written extensively on this subject in both scholarly and popular genres.[2] The title *The Mathematical Universe*, by William Dunham,[3] from Muhlenberg College, provides an instructive draft on the nature of mathematics and its evolution, covering the works of some of the most credited mathematical personalities.

This communication doesn't cover the roles of many of the ardent pursuers, founders, and discoverers of mathematical ways and methods, such as the prolific ones from ancient Greece. However, some of them are discussed here against specific contexts. Neither does this text intend to extend a chronological development of mathematics as an academic field. *The Road to Reality*, by Roger Penrose, provides a thorough account on the unfolding of mathematical sciences, from the ancient thoughts to the present-day understanding.[4]

Here my intention is to go step-by-step over how mathematics, augmented by analytics of physics, proposes a map of uniformity that sponges up the perceptivities, first utilizing simple arithmetical concepts and then moving into the relatively complex ones, such as those that describe higher-dimensional space.

Utilizing the same schematic, we will then attempt to grasp the alignment of the parallel universes, through which the sentient elements also flap.

Mathematics Imaging Three-Dimensionality

Self-Constituting Molds

The mechanics of arithmetical exponentiation recapitulate the materialization of three-dimensionality in the engineering of the universe—why nature is structured in precisely three dimensions and not in two or four, and why it doesn't follow any other vectorial standards. Any number, except for 1, raised to the power of 2 (that is, squared) or 3 (that is, cubed) pictorially brings forth the molds of two-dimensional (2-D) and three-dimensional (3-D) factuality, respectively. The squares and cubes of every quantitative embodiment can be depicted graphically by homogenously arranging the blobs of the quantity in two- and three-dimensional arrays, respectively. In fact, the development of the 2-D, as well as 3-D, structure is a spontaneous outcome if we are to evenly arrange the quantitative dollops exhaustively (figure 2.4). For instance, 3^2 (9) arrays itself into a 2-D form, whereas 3^3 (27) whips itself up into a 3-D structure. Similarly, 4^2 (16) freely takes on a 2-D motif, whereas 4^3 (64) swimmingly assumes a 3-D shell (figure 2.4). The square and cube of a numerical quantity emphatically simulate the 2-D and 3-D expositions of the actual nature—the area and volume in the parade of the tangible world. The eloquent drills of exponentiation don't just stop at the power of three; the regimen of exponential magnification publishes a layout that accentuates the draft of the physical measure beyond the dimension of three. But let us first eyeball where the ordinal *one* stands in all of this.

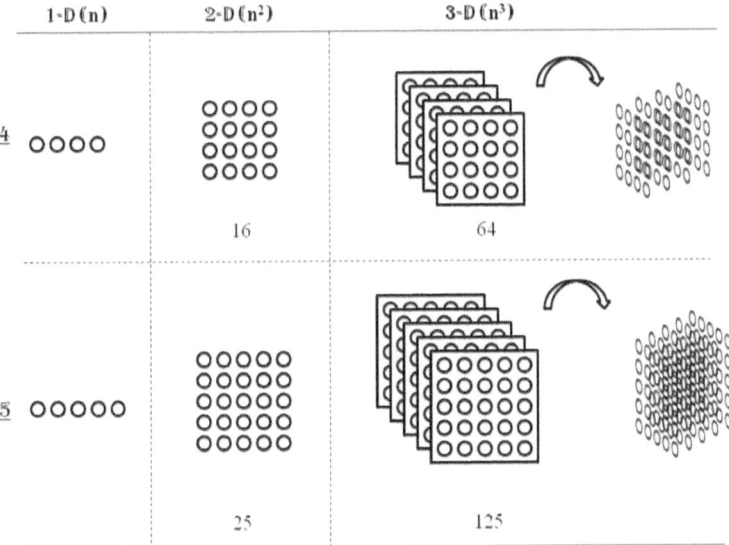

Figure 2.4. Exponentiations by two and three lead to the development of two-dimensional and three-dimensional configurations of reality.

Figure 2.4. (continued)

What is the timbre of the ordinal 1 in the sketch of reality? The number 1 does not multiply to itself, and it dresses identically in 1-D, 2-D, and 3-D presentations (since 1^2 and 1^3 both would still be 1). Also, the accent of 1 spells an uncleavable unit in which the multiplicities manifest as varied projections of the employed elements residing in that all-inclusive unit. Such an illustration, though, would show up only in the streaming of parallel universes and after accounting the observer. It is a little early to bring that backdrop into the phrasing here, but the reasoning is elemental in proposing the presence of a universal nondual mass–energy core, and then rationalizing the flapping of multifarious universes by it. The singleness of 1 clues the convergence of dimensions too into a single framework—again, the topic of concluding sections.

All ordinals, beyond the oneness of one, put on two-, three-, and higher-dimensional formats.

The authenticity of exponential displays in narrating the real cut becomes starker for the exponentiation beyond that of the third degree. And this is where we may be able to envision how the configuration of the higher dimension in reality shapes up. The appreciation of the perspective of a higher dimension is axial if we are to seize the flapping of multitudinous universes—the way quantum mechanics suggests—and if the subsumption of the sentient cadre within the absolute schema is to be contrived. It is exigent here to identify that there is a sleek and distinctive difference between the slants of "higher dimension" and "multidimensional"; whereas the previous points to parallel existences, the latter communicates the pulsation of space-time by the conformation of dimensions that are in continuum with each other, such as the image of space-time in eleven dimensions, the way the current insights in theoretical physics suggest.

An impeccable rendering comes into view when we exponentiate a count further the third degree. The packaging now, for exponents four and up, generates constructions that are procured not by the further annexation of the preceding three dimensions but by the duplication of the order of the initial 3-D cut (figure 2.5). Once the 3-D structure is acquired, any further vectorial enlargement musters only the accumulation of that crisp measure. For instance, 2^4 (16) propagates two of its own 3-D forms, and 3^4 (81) begets three of its own 3-D forms (figure 2.5). Similarly, 4^4 (256) gives four of the original 3-D shape and 5^4 (625) accords five of the beginning volume. Likewise, 4^5 (1,024) delivers sixteen of its own initial volume and 5^5 (3,125) breaks in twenty-five of the particular opening bulk.

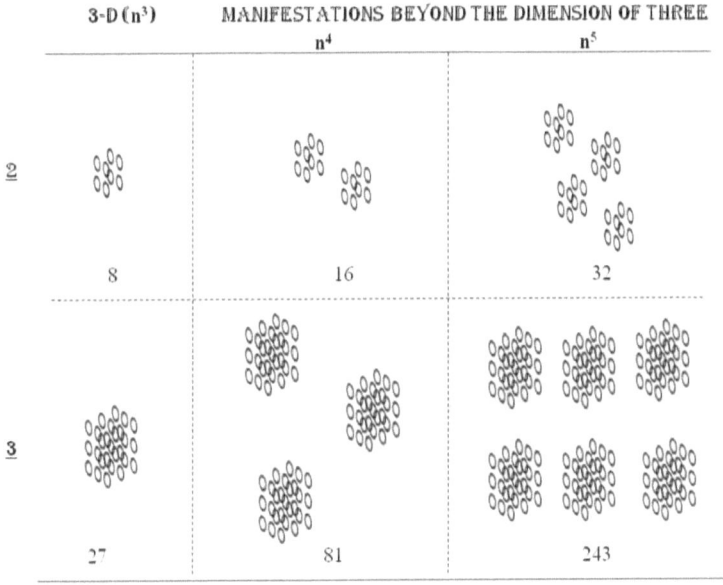

3-D (n^3) MANIFESTATIONS BEYOND THE DIMENSION OF THREE

n^4 n^5

| | 8 | 16 | 32 |

| | 27 | 81 | 243 |

Figure 2.5. The exponents beyond that of three only generate the copies of the original three-dimensional forms.

The dapper trims of the mathematical powers beyond the grade of three suggest a translucency in the proliferation of dimensions in the attainment of the physical world—that the annexations beyond the expanse of 3 do not take form, a fact of which we are doubly sure from the way the universe strikes the eye. Indeed, this does not legitimize that there aren't any additional ambits seeping into space-time, for we catch the fabrication of the universe only by the agency of physical forces, and thus, if the real picture is skewed, we won't know. For now, nonetheless, taking into account time as an independent dimension (the continuum in Einstein's context), these mathematical demeanors in a straightforward tone demystify how the higher dimensions can be crystallized parallel to the spatial 3-D bends, separated by the intervals of time. Howbeit, we will get back to all of this.

Cross-examining the preceding exposition may reveal yet another meaningful subtlety. Picturing it differently, the deduction suggests that two three-dimensional forms cannot be coalesced. For example, 3^3 (27) is a three-dimensional format that cannot be coalesced with another 27, the original bulk, or any numbers of the starting 3-D mold, to breed a single uniform, cleanly outlined structure. This holds true for any ordinal. The nonconjoinability of the two 3-D configurations furnishes, rather, suggestive hindsight: the discreteness that we are designed to perceive expresses in three dimensions only, and the ambience of the empty space by which the entities become apparent is the same by which the higher dimension takes shape as well (figure 2.6).

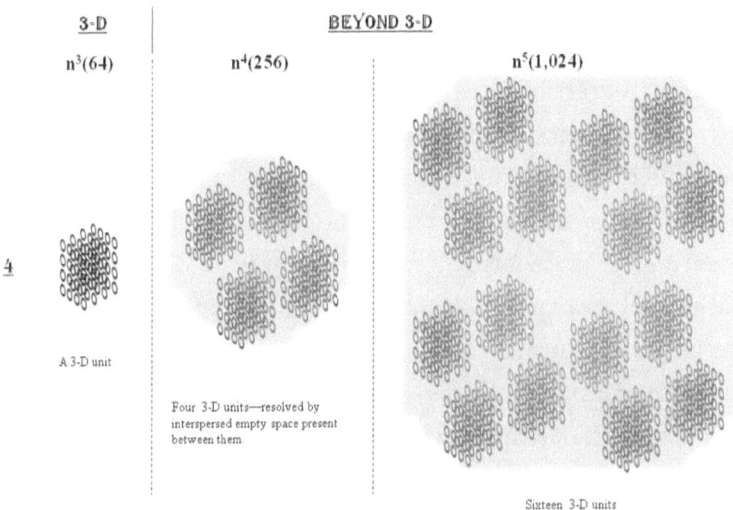

3-D	BEYOND 3-D

$n^3(64)$ $n^4(256)$ $n^5(1,024)$

4

A 3-D unit

Four 3-D units—resolved by interspersed empty space present between them.

Sixteen 3-D units

Figure 2.6. The dimensions beyond that of three, too, become audible only by the accompaniment of the empty space, spread alongside the show of embodiments.

Why the Happenstance of Three Dimensions?

Why is the universe three-dimensional? Why are we in tune with three dimensions? Why isn't another standard of dimensionality a natural outcome? The number of dimensions that sweeps into space-time has to do with how the dots of matter are connected in the codification of the all-inclusive prescript rather than with happenstance.

An entity in the tapestry of space-time remains interwoven with the rest of the entities, fully under the clout of the universal forces. The highly resolved image is imparted in quantum mechanical modes that we will meet in subsequent chapters. At any given point in time, thus, a blob of matter—say, a cosmic body—bunks only in relation to the rest of the matter in the terrain of the all-out compass. The number of dimensions that conform to the structuring of this built-in relatedness that couples the potencies of the fundamental forces harmoniously professedly are three.

The progression of space-time not only entails positioning in correspondence with the all-embracing physical laws it also concerns the fluent succession of continual motions under the sponsorship of the same forces. More of this will be covered after seeing the mathematical disposition of the circle in chapter 4. The geometrical motif that conforms to both the conditions—the relative positioning and continual progression under the uniform influence of universal forces—is the 3-D contour of the spherical mold. Suggestively, the way the cosmic arena broadcasts as well, it is only the spherical geometry that can exist as well as continually and frictionlessly move in a scheme where the exhaustive concordance is the first principle. Thus the spherical cast of the three dimensions apparently brings about the pitch-perfect geometry that is demanded by the fabric of space-time. The elaboration on the geometry of sphere in relation to reality is given in the section "The Perfect Pitch of Curve in Orchestrating the Concordance of Space-Time" in chapter 4, after examining the deep-seated essence of mathematical curve.

Let us now get back to the gist of this chapter—correlating elemental mathematics with the way the universe appears to the eye.

Fermat's Last Theorem: An Enigma, or Not

For its blunt accuracy and transparency, even though we didn't have a valid proof at the time it was stated, Fermat's last theorem became a cliché mathematical citation, appearing regularly in didactic and popular genres alike.[5,6] The statement is elegantly simple, but the meaning conveyed is both sharp and profound. Drafted by a French mathematician, Pierre de Fermat, in the year 1637, it states,

$$x^n + y^n \neq z^n$$

where n is the exponent of 3 and up. The phrasing tells us that the sum of two exponentiations cannot give rise to an exponentiated entirety for the powers of three and up. For example, 3^2 plus 4^2 structures into 5^2, but 3^3 plus 4^3, in accordance with Fermat's theorem, does not evolve into an entirety of x^3—3-D-fold. Fermat's equation applies for any numerical grade—in fact, tellingly, for any digital combination—as long as the power is 3 or higher. In the core of this premise firmly rests the efficacy of the aforementioned aphorism that the 2 of the three-dimensional units cannot be coalesced into an entire unit of three dimensions. Fermat's theorem is perched on the eminent framework of the Pythagorean theorem. Named after the Greek philosopher and mathematician Pythagoras (ca. 570–497 B.C.), it declares,

$$x^2 + y^2 = z^2$$

The sum of the squares of two numbers can be a square of another number. Specifically, the Pythagorean theorem stated that among three "square" areas, each perched along the hypotenuse, adjacent and opposite, respectively, of a right-angled triangle, the area along the hypotenuse is the sum of the areas that lie snug along the adjacent and opposite. The synopsis identifies the equation shown above. Only a set of combinations—commonly known as Pythagorean triplets—bestows a clean-cut dispatch of the indicative proposition. The combination 3, 4 and 5 is one of the most frequently seen Pythagorean triplets. The equation is effective only when two, two-dimensional forms can be summed into a single two-dimensional appearance. For example, 3^2 plus 4^2 can be arranged into a single two-dimensional state; the result would be 5^2 (figure 2.7).

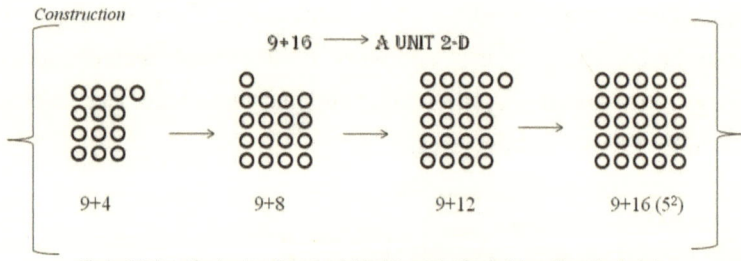

$$\boxed{x^2 + y^2 = z^2}$$

Figure 2.7. Summing two two-dimensional objects to construct a single two-dimensional whole.

Doing the same with the power of 3 does not work. For example, adding 3^3 (27) to 4^3 (64) fails to administer a single three-dimensional wrap. The numbers run short before the build of the clearly outlined 3-D form signs in (figure 2.8). We can try this with yet another Pythagorean triplet—say, 5, 12, and 13. The amount 12^2 can be arranged around 5^2 to generate a single two-dimensional embodiment, 13^2, but we cannot align 12^3 (1,728 articles) around 5^3 (625 articles) to have a uniform three-dimensional expression. The numbers again run short before the enclosure of a well-defined object takes form.

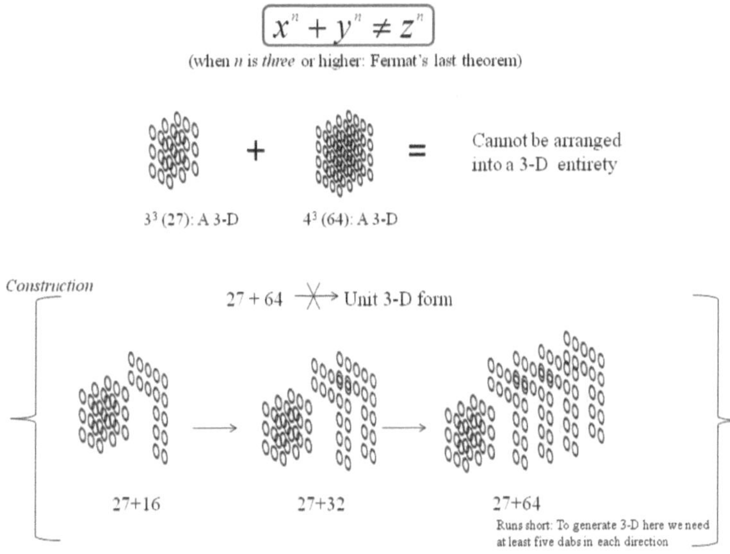

$$\boxed{x^{n} + y^{n} \neq z^{n}}$$

(when n is *three* or higher: Fermat's last theorem)

+ = Cannot be arranged into a 3-D entirety

3^{3} (27): A 3-D 4^{3} (64): A 3-D

Construction

27 + 64 $\longrightarrow\!\!\!\!\!\times\!\!\!\!\!\longrightarrow$ Unit 3-D form

$\longrightarrow$ $\longrightarrow$

27+16   27+32   27+64

Runs short: To generate 3-D here we need
at least five dabs in each direction

Figure 2.8. The summing of two 3-D units

The kinship between the mathematical speech and the actual three-dimensionality in the universe—that the spatial extents of distinct objects do not grow beyond that of three—makes Fermat's last theorem self-evident. Providing a proof in formal technicalities, however, is yet another affair and is critical within the domain of the subject. Fermat's last theorem was mathematically proven to be authentic by a British mathematician, Andrew Wiles of Oxford University, in 1994, about three and a half centuries after it was proposed.

The pictorial depiction of the power augmentation exposes a foundational feature that the 3-D configurations of either the same or two different numbers do not mingle to furnish a single solidarity of 3-D accent. On a simpler note, and stating it differently, it is impossible to build a cube that has twice the volume of a given cube. This was known to ancient Greeks as "duplicating the cube problem" and was later proved by a French mathematician Pierre Wantzel.

Then what comes to be in the cases of a higher order of amplifications, such as those by the exponent degrees of 4, 5, and 6—x^4, x^5, and x^6? Nothing can be coalesced once the three-dimensional form gets in.

In a little while we will gather that up-to-date and paramount understandings in the field of quantum research suggest that in the orderliness of space-time there careen eleven dimensions instead of the obvious three. I will try to scoop up those deflections of reality as well when I get into the subject with the backdrop of additional relevant information. Here the revelatory assertions of power expansions highlight that in the nature of space-time further to the extension of three lies the continuation of empty space, separated by which, thus, materializes the matrix of higher dimension.

The three-dimensional entities, as discrete spaces, stand out in consequence to the empty space that weaves between them. The synchronicity between the elements and the observer—the rhythms of mass–energy alongside the nestled watcher—is analogous to the scene of mathematical entities spreading in concurrence to the presence of the empty space (figure 2.6). The mathematical empty space reflects the very same observer state that is affixed to the actuality of all that crystallizes.

The form 4^3 is a 3-D unit, and 4^4 then becomes the four copies of that unit, where 4^5 comes to be sixteen imitations of the same articulation, all expressed concomitant to the disclosure of the empty space, the continuum of the same gap that identified the first three-dimensional appearance (figure 2.6).

What Is Mathematical Gap?

Although the pictorial depictions of algorithmic gap and the compass of the observer of certainty are equivalent in the sense of appearance, they do not dispense the same meaning; that is, the blank extents that we encounter around us and in the vastness of the cosmos would not equate to the mathematical gaps. According to the Heisenberg uncertainty principle—named after German theoretical physicist Werner Heisenberg, who gave the basic introduction of this doctrine—we cannot simultaneously determine the position and the momentum of a particle with equal precisions. The more accurately one physical property is determined, the less accurately the other physical property will be reflected. Inferring this discovery in relation to the quantum field theory led to the scenario in which the appearance of blank space became a disallowed feature. This means that, in actuality, there is no blank space present (between physical entities), whether at quantum or at cosmic scales.

The scientific rationality of the presence of open space that we perceive all around us is that although we see the objects of the universe as independent disconnected entities, in reality they all remain unified to the perceiver through the universal forces; the forces of electromagnetism and gravity are the underlying reasons for all that we perceive with the naked eye. So the blank space between eye and object is not there in the real sense, when we account both for the core of relativity and for the exertion of physical laws through which the matter beats, or see the picture the way quantum mechanics presents.

The most advanced models to understand the texture of spacetime, such as those set out by the string theory, rationally do away with the rather factitiously molded blank spaces; the mass–energy and the comingled universal forces are exhibited by the vibrating modes of the strings. Therefore the mass–energy and the forces through which it exists are amalgamated, yielding an all-encompassed, self-permeated space-time; there isn't any barren area exposed between the two objects. The perceived empty space between the two objects in reality is occupied by the swell of forces between the two entities, no matter how far apart the embodiments lie from each other. Therefore the blank extents of the universe that strike the eye de facto are smoothly tailored stretches of matter and forces in the continuum of space-time.

The estate of witnesser that camps concatenated with physical reality, therefore, in that context, is different from the blankness that we sense around us and in the cosmos. The sweep of the witnesser lodges independently and thus is resilient to the space-time undulations. The mathematical empty space, however, demonstratively equates to this propertyless state of the witnesser; the digital objects too dwell concatenated to the gap that extends alongside their casements. In the light of the conceptual role the empty space plays in defining the power set (in the set theory) toward understanding the space-time warp that occurs in reality, the directness of the mathematical gap in picturing the reality becomes even more self-evident. We will see this in chapter 7.

The Messages of Numerical Spaces

The numbers as mathematical entities sprucely imprint the real space; the numerals as *is* or their acquired structures via mathematical operations, whether simple arithmetical or advanced abstractive, deftly highlight materialistic lineaments. Along the same lines, the "spacing" set forth by different number types evidently simulates different aspects of the real space.

Irrational numbers, though, are denser for a given space, compared to whole and rational numbers, yet still are exhibited by the interlaced gap between them. Irrational numbers, however, lack discreteness, and therefore, in that context, they occupy an intuitively different indefinite turf, compared to the rational and whole numbers (figure 2.3). To that end, the irrational numbers accentuate an aspect of the real space that is not discernible by rational and whole numbers (more on this in the following chapters).

The impalpable nature of the irrational numbers means that they constitute their own infinite expansion, which is independent of infinities of the whole and rational numbers. In his exploration to probe the temper of infinity, Georg Cantor showed that although incalculability of the irrationals places them at a higher level of infinity, the uncountable set itself can be further aggrandized into an even larger lot of infinity—a process that sets off the deposit of endless arrays of infinities, the series that he tagged with the Hebrew character aleph, ℵ, where a numeral subscript that followed aleph marked the level of infinity in the series. This shaped the coming of the transfinite series, noted earlier.

Irrational numbers seemingly expose certain quirky, and yet justifiable, features of space-time, as well as the peculiarities of our comprehensions that are attuned accordingly within it—the topic of the next three chapters. The mathematical irrationality featured space-time, encompassing our in-tuned comprehensions, evidently relates to the universe that we grasp from the theories in physics. All along I will attempt to forge the delicate connection between the revealing nature of mathematical concepts and the discoveries that sprout in physical sciences.

CHAPTER 3

The Voice of Transcendental Numbers

The Numerical Expression Hinting toward the Boundless Cast of Matter

The Lens of Transcendence

Literally, the word transcendence insinuates surpassing usual limits, an attribute infused with abstruseness that cannot be straightforwardly defined. At the physical level, *transcendental* proclaims an inability to be discretized, or that which cannot be presented as a freestanding whole. In mathematics, "transcendental" means the same—a disposition that cannot be declared by a unitized numeric value. A transcendental number, the existence of which was first shown by French mathematician, Joseph Liouville in 1844, is a type of irrational number that not only defines an unlimited decimal expansion devoid of a numeric pattern but heralds an execution that is nonalgebraic as well. It is nonalgebraic because a transcendental number cannot be fitted into a solvable algebraic equation, or it cannot be expressed in a form of an algebraic equation. Phrased lucidly, transcendental numbers are nonquantifiable.

The ratios 2/3 and 12/7 bring about numeric patterns in their unlimited decimal expansion:

$$2/3 = 0.66666666666666666666...$$

$$12/7 = 1.71428571428571428571428571428571428...$$

The two preceding numbers expand infinitely, tagged with specific number patterns: 6s in the first example and a repeat pattern of 714285 in the second example. Both of the numbers, therefore, are rational numbers. The square root of two, $\sqrt{2}$, is an example of an irrational number, as it lays down a limitless decimal expansion that does not host a rolling numeric pattern; however, it is not a transcendental number, because it is algebraic—it can be expressed by an algebraic equation ($x^2 - 2 = 0$, where the $x = \sqrt{2}$). Thus, it is quantifiable, depicting a finite quantity that can be expressed via a mathematical statement.

Pi (π), the ratio of a circle's circumference to its diameter (figure 3.1), has lured mathematicians and logicians for thousands of years, for its deportment of circularity not only defines the underlying framework of immense disciplines in mathematics and physics but also extends to suggest how the universe itself is structured. Calculus, trigonometry, all of the avant-garde algebraic formulations, mechanics, electromagnetism, cosmology, relativity, and all of the pioneering constructs of space-time are just a few of the obvious branches in which the nature of π participates indispensably. In a deep sense, there is no scope of mathematics or physics where the insights from π aren't of utility. Pi (π) was proved to be a transcendental number by German mathematician Carl Louis Ferdinand von Lindemann in 1882. It unveils the uninhibited decimal expansion that is devoid of any rolling pattern, and it is nonalgebraic as well, emblematizing a nonquantifiable state.

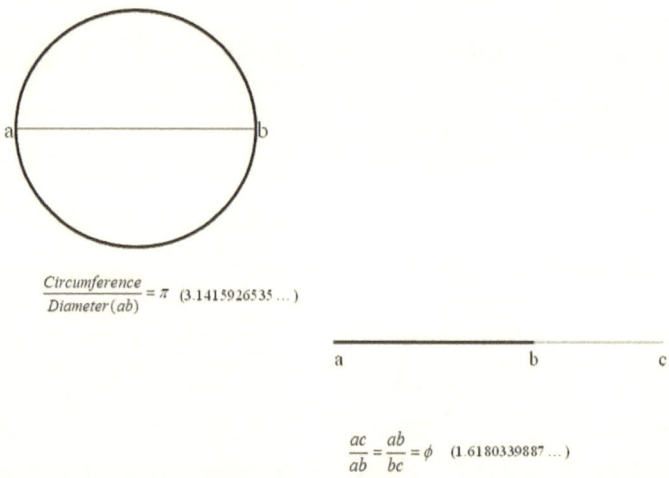

$$\frac{Circumference}{Diameter\,(ab)} = \pi \quad (3.1415926535\ldots)$$

$$\frac{ac}{ab} = \frac{ab}{bc} = \phi \quad (1.6180339887\ldots)$$

Figure 3.1. The transcendental numbers Pi (π) and Phi (φ).

Its nonconvergence has been computed to the extent of 10 trillion decimal digits.[1] In the following, π is stretched to its nine hundredth decimal place.

3.14159265358979323846264338327950288419716939937510
58209749445923078164062862089986280348253421170679
82148086513282306647093844609550582231725359408128
48111745028410270193852110555964462294895493038196
44288109756659334461284756482337867831652712019091
45648566923460348610454326648213393607260249141273
72458700660631558817488152092096282925409171536436
78925903600113305305488204665213841469519415116094
33057270365759591953092186117381932611793105118548
07446237996274956735188575272489122793818301194912
98336733624406566430860213949463952247371907021798
60943702770539217176293176752384674818467669405132
00056812714526356082778577134275778960917363717872
14684409012249534301465495853710507922796892589235
42019956112129021960864034418159813629774771309960
51870721134999999837297804995105973173281609631859
50244594553469083026425223082533446850352619311881
71010003137838752886587533208381420617177669147303

Pi expanded for ten thousand decimal digits can be seen at the university of Utah website.[2]

Here the cause of the seemingly mesmerizing uninhibited decimal expansion lies in the relationship via which the straight and the curve prevail. The number embellished transcendence delicately hymns the nature of the ultimate geometry, that which tailors the continuum of space-time. Before attempting to conjoin the mathematical transcendence with the nature of space-time, we need to examine how mathematical transcendence actually comes about. But, even before that, let us first have a peek at the flourished pre-eminence of mathematical transcendence in the actuality of nature and the universe.

The Word of Transcendence in Nature: The Golden Ratio (ϕ)

Euclid—Euclid of Alexandria (ca. 300 B.C.), who pioneered the field of geometry and is known for his treatise *Elements*—through the geometrical imports brought the pitch of a transcendence to our attention that later unfolded as one of the most awe-inspiring, as well as graphical, inscriptions of mathematical form in the build of the natural world. This transcendence is that of phi (ϕ), also known as the golden ratio. The golden ratio results when the ratio of the sum of two quantities relative to the larger quantity is the same as the larger quantity relative to the smaller one (figure 3.1). The appellation "golden ratio" itself is testimony to its pervasive presence in all aspects of the worldly layout.

This ratio appears in leaf, floral petal, and seed head arrangements; it shines in the structuring of shell spirals and in human body proportions, including the face, hand, and ear; it is declared in the layout of fractals that abound in natural phenomena; it hums in the sway of musical notes; it hovers in creative illustrations, architecture, and geometrical shapes that appeal to the human mind; it permeates the anatomy of natural objects, from the quantum to the cosmic scales—to name just a few instances of its copiously homogenous presence in all that seems to exist physically and is perceived introspectively. The title *The Golden Ratio*, by Mario Livio, offers meticulous coverage of the credibility and the historical background of this effusive measurement. The web pages developed by Ron Knott at the University of Surrey provide an illustrative account on this ratio, packed with fitting selections of coherent examples.[3]

The decimal expansion of the Golden Ratio (ϕ) reaching the one thousandth decimal place is shown in the following:

1·6180339887498948482045868343656381177203091798057 6
2862135448622705260462818902449707207204189391137 4
8475408807538689175212663386222353693179318006076 6
7263544333890865959395829056383226613199282902678 8
0675208766892501711696207032221043216269548626296 3
1361443814975870122034080588795445474924618569536 4
8644492410443207713449470495658467885098743394422 1
2544877066478091588460749988712400765217057517978 8
3416625624940758906970400028121042762177111777805 3
1531714101170466659914669798731761356006708748071 0
1317952368942752194843530567830022878569978297783 4
7845878228911097625003026961561700250464338243776 4
8610283831268330372429267526311653392473167111211 5
8818638513316203840052221657912866752946549068113 1
7159934323597349498509040947621322298101726107059 6
1164562990981629055520852479035240602017279974717 5
3427775927786256194320827505131218156285512224809 3
9471234145170223735805772786160086883829523045926 4
7878017889921990270776903895321968198615143780314 9
9741106926088674296226757560523172777520353613936 2

The Coming About of the Mathematical Transcendence

Like any other transcendental number, the golden ratio expands infinitely and does not host a predictable pattern of rolling numerals, or is nonquantifiable. The abysmal gesture that is lucidly illustrated in mathematical terms perhaps furnishes clues on the way universe manifests and the role the attribute of transcendence might be playing in optimizing the universal rhythm. Mathematically, the absence of *definitive* numeral entities—those between which we are trying to find the ratio, the x and y in x/y—is the very underlying cause that effectuates the surpassing behavior. The shadow of this "inexactness" rests on the configuration of interdependency, whether we are talking of mathematical transcendence or the one in the physical plane. The two quantities, vis-à-vis attributes, which exist only interdependently and cannot be dissociated would conjunctionally evoke the intonation of transcendence. I will get back to the relative existence of the two attributes, especially in context with the subsistence of the curve in relation to the straight, the differentia that engenders pi (π). Let us first glimpse what the image of transcendence means in mathematical terms.

The ratio 34/3 sets out 11.3333333 ..., with unending decimal repetitions of 3, stating that the denominator cannot split the numerator unambiguously; there will always be a leftover, and that leftover will endlessly recur in the decimal expansion. This is an example of a rational number, as we can pretty much figure out how it materializes—the culmination can be rationalized, both mathematically and conceptually. How does this ratio compare with the one that is transcendental—that is, the ratio that is prolonged in the absence of any numerical motif and is nonalgebraic? In fact, such an abysmal lengthening and the number being nonalgebraic are just two facets of the very same gesture of transcendence.

The mathematical transcendence might be better ascertained by looking into its underlying cause rather than examining the effectuated boundlessness of its spread. The ratio 6/2 begets 3. The numerator (6) evenly splits into the denominator (2), generating 3 units of 2, the denominator. Invoking transcendence mathematically, however, is engagingly perplexing because it obligates finding two of those mathematical quantities that can beget such a transcendental phenomenon. The seemingly captivating mathematical task is insurmountable plainly because mathematical transcendence purports change in perpetuity, including those initial characters by which the ratio was set forth. "Transcendental" by definition means "never to converge," so the two quirks always manifest associatively. Both will continually appear alongside the act of measurement.

The enumerative transcendence in the dialogue of mathematics can be established by utilizing trigonometric functions that tie the character of the curve into the measure of the flat—and the emergence of transcendence has to do with how the curve exists in relation to the straight.

The relevance of the curve becomes increasingly apparent once we have taken a glimpse into the intriguing nature of pi (π), the transcendence that hums in every core of existence—the theme that I strive to broaden in the next two chapters. The mathematical transcendence expressed by pi (π), and so by the curve, sheds light on the way actual transcendence in nature and the universe sweeps, and correspondingly, by ascertaining the primal nature of the curve, the compounding component in the geometry of the universe, we may be able to unweave certain mysteries that evidently ignite the absolute transcendence—that is, space-time.

Here one such mysterious factuality warrants mentioning: according to the thriving interpretations from quantum studies, the most basic ingredients of the universe lie spatially extended, smeared into space-time, rather than having discrete etched structures. The scientific understanding dictated by quantum mechanics, therefore, depicts the existence of matter (the mass–energy forms) abstractly, swept through all the space dimensions, governed by the surrounding forces. Thus the embodiments of space-time are pictured as the shades of inexhaustible passages. And this encompasses the living entities as well.

In view to the way we sense ourselves and the cosmos, these findings are almost mind bending, albeit they highlight the deeper level of the reality, whether seen quantum mechanically or realized subjectively. What appears to be outwardly magical, however, can be straightforwardly equated in view of the mathematical transcendence, which not only simulates the tones of a bleary and indefinite description of reality but also draws attention to how the relativistic space-time comes about—more on all of this in the following chapter. Right now, let us navigate a little more through the basal aspects of the uncanny transcendence and the various ways in which it unites mathematics and the universe into the same game.

An unfettered decimal expansion, free of any numeric pattern, signifies constant change. A ratio of two natural numbers, although it can bring on an unlimited decimal expansion, will always be accompanied by mathematical signatures and therefore cannot assert transcendence. Transcendence manifests when the two initial peculiarities constantly change as we continue to assess the ratio, and mathematics is the way by which we discover that the phenomenon of transcendence pervades the systematic order of the real world. The topic of numerical transcendence to delineate the nature of existence has inspired mathematicians, philosophers, logicians, artists, scientists, and even psychotherapists for thousands of years. The ratio of 22/7 is 3.14285714285714285 ... The number is close to pi (π) (see above). But π loses the pattern soon after a few decimal places—instead of 3.1428571 ..., it goes 3.14159265358979323846 ... toward a completely different unknown way. How do we infer such a mathematical behavior? In a mathematical dictum, such a bearing announces that although the ratio was initiated through somewhat discrete mathematical entities (somewhere close to 22/7), these quantities themselves kept changing interrelatively as the ratio continually set in—although at the technical level, the duress behind this marvel, as we have noted, is the mathematical sewing of the trait of the curve onto the outright ambits of the straight. Enlivening wonder, such mystical quirks seemingly imbue every iota of the universe, from the macrocosmic to the intellective fields. To savor the aura behind their genuineness is itself phenomenal.

The Digits Announcing the State of Transcendence in Cosmic and Psychic Amplitudes

Ever since the intuitive realization of the rendering of the golden ratio in geometrical shapes—during the time of ancient Greeks, when it appears that the golden ratio was employed in architectural and inventive creativities. The detected examples of this ratio in nature, in the universe, and in the works of human pursuits have cumulated by leaps and bounds. There is absolutely no end to it—structural, anatomical, algorithmic, physiological, atmospheric, meteorological, economical, pictorial, musical, phrenic, you name it—the Golden Ratio crops up in all quarters. The presence of the golden ratio (ϕ) envelops structural aspects of plants, animals and nature, from cosmic to quantum planes; the ratio defines behavioral trends, such as growth and movement (natural and economic); it soaks up the emotional bearings, such as those that identify liking and taste, and the creative deportment that establishes the artistic pursuits.

Some of the most frequently cited examples are as follows: the arrangement of seeds in the sunflower and daisy reflect the golden ratio; the ratio is seen in the growth patterns of many plants; the ratio spreads in a range of intrinsic spiral builds, such as those of seashells, flowers, seeds, vegetables and fruits, DNA, and galaxies (figure 3.2); it drums in the way quantum field behaves;[4] the human shell (figure 3.3) in totality, as well as the anatomy of face, ear, and fingers, totes the scale of the golden ratio in its composition; apart from the anticipated skeletal occurrences, the ratio sprouts in population growth, weather patterns, and stock market shifts. As I mentioned earlier, the site by Ron Knott, University of Surrey, has the most integral examples descriptively covered and mathematically mapped. The sharp connectivity between this ratio and the framework of nature is excellently depicted in a video by Cristóbal Vila.[5]

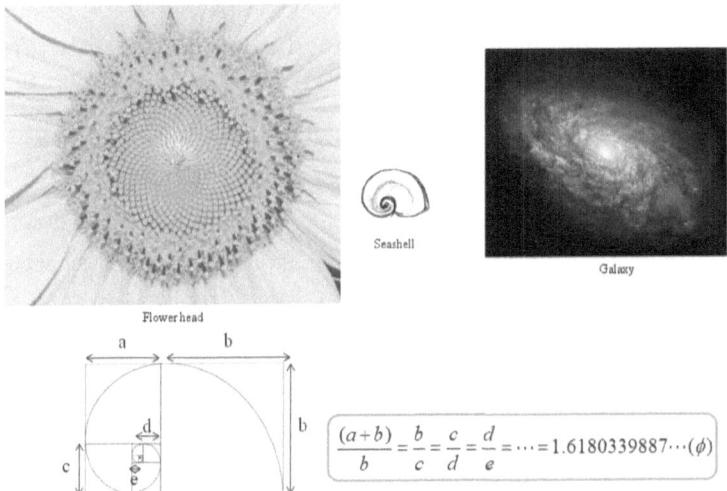

$$\frac{(a+b)}{b} = \frac{b}{c} = \frac{c}{d} = \frac{d}{e} = \cdots = 1.6180339887\cdots(\phi)$$

Figure 3.2. The assimilation of the golden rectangle in naturally occurring spirals.

The galaxy image is downloaded from ORIN (Great Images In NASA)

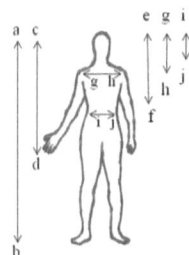

$$\frac{ab}{cd} = \frac{cd}{ef} = \frac{ef}{gh} = \frac{gh}{ij} = 1.6180339887\ldots(\phi)$$

Figure 3.3. The golden ratio in anatomy.

Even more extraordinary is the fact that the golden section (the proportion that displays the golden ratio) equally infiltrates those articles that are designed by humans (knowingly or unknowingly). The most lustrous architectures, such as the ancient Greek structures of the Parthenon (figure 3.4) and the Porch of Maidens; the great pyramid of Giza (the ratio of slant height to pyramid half-base gives the golden ratio to an accuracy to the fourth decimal place; figure 3.5); Notre Dame; and Chartres Cathedral all display the golden ratio in their enriched appeal, in various ways. The framework of the Parthenon embraces the cuts of golden rectangles throughout its build. The golden rectangle comprises dimensions such that if a square is removed from within its structure, the leftover rectangle will have the same length-to-width ratio as that of the original rectangle (figure 3.4). Naturally occurring spirals, such as in nautilus shells or galaxies, wind along the measurement of the golden rectangle. The United Nations building is an example of modern architecture in which the festoon of the golden ratio is apparent.

The craft of the *Mona Lisa* painting (figure 3.5), which was created by the distinguished painter, architect, mathematician, scientist, and engineer Leonardo da Vinci (ca. 1503), is considered to be the one of the finest artistic creations, and its flawless appeal in part seems to be rooted in the portraiture's blueprint, which deftly flares the ornamentations of the Golden Ratio. For example, the streak of the golden ratio has been claimed between the measures of the neck base to the forehead top, relative to the neck base to the pupil center; between the measures of the chin bottom to the nose bottom, in respect to the chin bottom to the lip bottom; in head length compared to head width. And it is not just in the outline of the face; the layer of the golden ratio is recoded in overall portrait measure as well. For example, the distance between the right fingertip and the top of the forehead sets the golden ratio with a distance between the right fingertip and the neck base.

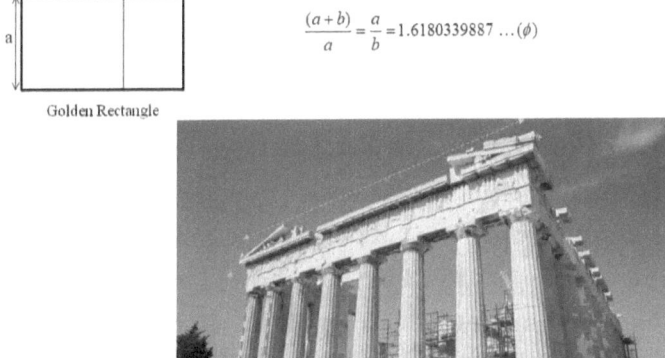

$$\frac{(a+b)}{a} = \frac{a}{b} = 1.6180339887\ldots(\phi)$$

Golden Rectangle

Figure 3.4. The occurrence of golden rectangle in the architecture of the Parthenon. The golden rectangle appears in different ways. Some are shown with dashed lines

Mona Lisa

Pyramid of Giza

height   slant height (x)

½ base (y)

Pyramid base

$$\frac{x}{y} = 1.61804\ldots$$

Figure 3.5. Golden ratio surfaces in brilliant artistic aptitudes.

The polyhedral depictions in *De divina proportione*[6] ("On divine proportion," a reference to the golden proportion)—the notable book on the demonstrative golden ratio by Luca Pacioli (1509)—seems to indicate that Leonardo da Vinci knowingly incorporated the golden ratio in his works. (See the explanatory online article "Leonardo da Vinci's Polyhedra" by George W. Hart.) *The Vitruvian Man*, a sketch by Leonardo da Vinci that features the human body dimensions up against geometrical shapes, also announces the presence of the golden ratio in more than one way. Leonardo da Vinci is known for his brilliance in painting, sculpture, architecture, music, and science, but, to the amazement of many, he also had an apt mathematical bent of mind. The lucid grasp of such diverse faculties conceivably relates to his imaginative endowment, leading to wondrous fruitions—that perhaps required optimized blending between the faculties of science and art. Less comprehended is why these works of art bring accentuated felicity in the tonality of the human mind.

Evidently there has been an eloquent connection between aesthetic appeal—whether in architecture, painting, sketch, or simple geometrical objects—and the echo of the golden ratio. For example (as I have noted from the title *The Golden Ratio* by Mario Livio), in a set of rectangles, the rectangle that is reinforced by the measurements of the golden ratio apparently would be the most engaging. One would instinctively feel in tune with that particular rectangle against the range of rectangles that do not reflect the golden ratio (even if one does not know beforehand which rectangle exhibits the golden ratio or does not have any idea what the golden ratio even is).

Musical compositions and sonatas have been acknowledged to reflect the golden ratio as well. Musical scales and frequencies, and even the divisions in the musical instruments, are based on the cut of the golden ratio. Symphonies of the highest caliber, such as those by Mozart, Beethoven, and Bach, inseparably employ the frequencies that conform to the lineament of the golden ratio.

The delicately dispersed cornucopia of the golden ratio both at the physical and cognitive levels articulately points to the cast of the same constructional theme behind the configuration of the universe and the manifestation of the mental cast. The overarching presence of the golden ratio in nature and the universe is further compounded by its presence in our intelligent choices too; we perchance happen to be mathematically in tune with the way the universe is, or vice versa.

The most acclaimed artists, such as Michelangelo, Raphael, and Leonardo da Vinci, might have knowingly crafted their work around the scale of the golden ratios, although this intuitively feels unlikely, as creativity is rather akin to more of a natural flow than a calculative production. In any case, the bearing that both physical and mental aspects of the universe fall under the same umbrella of the mathematical texture may point to the roots of dexterity that imbue these world-class artists. And it is just to say that these virtuosos clothed in a fine-tuned natural sense perhaps would have instinctively produced ravishing pieces of abstractions that, in the accent of mathematics, would have consequently transmitted the ribbons of the golden ratio.

The more we hunt, the more we bump into the ubiquitous instances of this ratio in static as well as functional aspects of the universe. It is as if both the universe and the mind are conditionally tuned to 1·618033988749894848204586834365 ...

Another Angle of the Golden Ratio: The Fibonacci Sequence

Sleekly asserting the same overarching correlation between the universe and mind that was brought to light by the voice of the golden ratio, yet bearing a purely unconnected origin, is the hammer of the Fibonacci sequence, named after Italian mathematician Leonardo of Pisa, or simply Fibonacci, who in his book *Liber Abaci* ("The Book of Abacus," published in 1202) introduced this series to Western Europe. The sequence resulted in the solution to a formulated population growth problem, composed by Fibonacci. The problem asks for number of rabbit pairs produced each month, starting from a single pair of rabbits. The conditional breeding defined in the problem is such that each pair produces a baby pair comprising a male and a female every month, where the baby pair starts reproducing the following month. Laying out the solution as the total number of baby pairs occurring subsequently each month starting from the original pair results in a number sequence now popularly known as the Fibonacci sequence:

0, 1, 1, 2, 3, 5, 8, 13, 21, 34, 55, 89, 144, 233, 377, 610, 987, 1597, 2584, 4181, 6765 ...

The preceding series shows the number of baby pairs serially appearing every month, starting from the first month. There will be one baby pair in the second month. In the third month, this baby pair matures to adulthood, while the original adult pair gives rise to another baby pair. Thus, in the third month there still will be only a single baby pair. In the fourth month, however, there will be two baby pairs—one pair from the matured pair and the other from the original. The same way, by the fifth month there will be three baby pairs—two pairs from the two adult pairs and an additional pair from the newly matured pair. And so on. The sequence appears whether we count the number of baby pairs, the number of adult pairs, or the total number of rabbit pairs.

The seemingly simplistic appearance of a number array cryptically bore crisp order, which is quite different from the actuality of a rabbit's breeding cycle, placing this series akin to the golden ratio in broadcasting the uniformity between the mathematics and the universe. The appearing sequence is not just some whimsical series of numbers; it is a layout of a structured behavior. Starting with the third in the series, every number is the sum of the two preceding numbers $(0 + 1 = 1; 1 + 1 = 2; 1 + 2 = 3; 2 + 3 = 5; 3 + 5 = 8; 5 + 8 = 13; 8 + 13 = 21; 13 + 21 = 34$, etc.$)$.

The most intriguing is this: as we climb up this sequence, the ratios of the two consequent numbers start to reflect the stamp of the golden ratio. As we go higher up this sequence, the ratio from the Fibonacci sequence gradually converges into the golden ratio (table 3.1). As with the golden ratio, the Fibonacci numbers are encountered all around, in the plant and animal kingdoms, in the amplitudes of the sea, in the mold of physique, in growth patterns, and in solar systems and galaxies. All in all, the encountered similitude between Fibonacci numbers and the Golden Ratio justifiably conveys their copresence throughout the universe. The Fibonacci sequence and the golden ratio sublimely reflect the same phenomenon. Their discoveries sprouting from distinctly different mathematical arenas to announce the very same precision of nature and universe, is truly revelatory, corroborating the well-assimilated conjectural view that mathematics as a subject is discovered rather than invented.

The Golden Ratio versus the Fibonacci Sequence: A Technical Variance

The marked semblance between the ratios from the Fibonacci series and the golden number, or ratio, is accompanied by their mathematical distinctiveness. The underlying contrast once again is rooted in the rational formalism of how these numbers come about. The Fibonacci series is generated utilizing a methodical problem that addresses discernible entities—the number of rabbit pairs—and thus deals with the whole numbers. Whole numbers are distinctively individualized objects, and therefore, as we have seen earlier, they cannot be combined to beget a number that is transcendental. This means that although the combination of two numbers from the Fibonacci series can engender a ratio that imparts an unending decimal expansion, they will nevertheless express a rolling pattern of mathematical nature (table 3.1). For example, the ratio 610/377 gives unlimited decimal expansion, with an 84-digit recurrence pattern. The ratio 1597/987 (table 3.1) provides a 138-digit recurrence pattern. If you feel investigative, you can take a peek at the nature of mathematical recurrences utilizing an online facility set up by Ron Knott of the University of Surrey.[7] Regardless of the systematic deviation from the golden ratio, the ratio emerging from the Fibonacci series nonetheless features the evenness between nature and mathematics, the conformability that is further extrapolated by the shimmer of the golden ratio.

Golden ratio (φ):

1.61803 39887 49894 84820 …

Fibonacci sequence:

0, 1, 1, 2, 3, 5, 8, 13, 21, 34, 55, 89, 144, 233, 377, 610, 987, 1597, 2584, 4181, 6765, 10946, 17711 …

Growing accuracy
with which the golden
ratio is echoed

x	y	The Ratio (y/x)	Repeat pattern
0	1	1	
1	1	1	
1	2	2	
2	3	1.5	
3	5	1.666666666666	
5	8	1.6	
8	13	1.625	
13	21	1.615384615384…	6-digit
34	55	1.617647058823…	16-digit
55	89	1.618181181818…	2-digit
89	144	1.617977528089…	44-digit
144	233	1.618055555555…	5s
233	377	1.618025751072…	232-digit
377	610	1.618037135271…	84-digit
610	987	1.618032786885…	60-digit
987	1597	1.618034447821…	138-digit
1597	2584	1.618033381340…	133-digit
2584	4181	1.618034055727…	144-digit
4181	6765	1.618033963166…	336-digit
6765	10946	1.618033998521…	10-digit
10946	17711	1.618033985017…	420-digit

The φ being 1.6180339887…

Table 3.1. The ratios from Fibonacci sequence mirror the golden ratio. The ratios between the first few successive digits of the Fibonacci sequence are shown. As we move forward along the Fibonacci series, the ratios between the adjacent digits more and more decisively exhibit the cast of golden ratio.

The Transcendence: The Underlying Space Perpetuity

As we will see in the succeeding chapters, the quality of transcendence is vital to nature and the universe, and to our existence within it. The show of mathematical transcendence in the scenery of the reality accentuates the framework of relativity that we make out from principles of physics, including Einstein's. Furthermore, the relativistic sea of space-time can discernibly accommodate views from quantum theories—such as the conception of the simultaneity of the observer and that which is observed, or the idea that what we perceive as discrete behind sight quivers as dispersed (and I will muse on these analytical outlooks in posterior sections)—by the torch of mathematical transcendence. It is the numerical transcendence that spotlights the enactment of the central principles uttering that there appears to be a solitary basal framework of an all-embracing space-time—the space-time that describes the conscious presence and the tracks of mind that it accompanies.

And it is crucial to acknowledge that the two transcendental gestures that enrich the universe, pi (π) and phi (ϕ), in essence radiate one and the same essence. At the constitutional level, both the deliveries of transcendence remain intrinsically interwoven. The transcendence given out in the form of the rather incognito phi (ϕ) architecturally rests on the transcendence that is seen in the form of pi (π). We will get back to this while attempting to understand pi (π)—the surpassing mannerism that dwells in all façades of existence, in amazing ways.

What does the ubiquitous presence of transcendental numbers tell us about existence, about our own stationing in the scopes of the ultimate space-time rhythms, and about the world that we create? How can we bring to terms the underlying mathematical transcendence of this physical world to the physics of all that we learn empirically? How do we construe a physical state, vis-à-vis a mathematical number, that conforms to the nonconvergence through the unfailing change?

The unlimited decimal expansion betokens concealed relativity, between the two quantities that initiated the ratio. This means that x is relative to y in x/y; the x is there because of the y. The variables x and y coexist; they do not exist independent of each other. A clean division of x/y results in an integer (or a converged decimal) because eventually y gets cancelled out. Envisage this: if y is the one that creates x, then from the window of the x the y will always be seen, no matter how small the x becomes. The y shows up as soon as the vantage point of x is available. And although a sort of pseudo-relativeness is hidden in any mathematical irrationality, transcendence is a step ahead, because it further bores perpetual changes, fitted through the banners of two interdependent attributes. The transcendence is an inkling of how the paradigm of mathematics mimics the temporal truth—the truth that embodies the essentialities of relativity and the décor of perpetual modifications. This is where the astonishing transcendental number pi (π) may provide us with esoteric clues.

The relativistic implementation of universe as cued by physics is the very same description of space-time (we will see these determinations later down the road) that the mathematical transcendence appears to rubber-stamp on its own.

The Elements of Mind in Tune with the Transcendence of the Universe

The drenching nonconvergence through all corners of the universe thus connotes that everything is relative to everything else. This is the understanding we have imbibed from modern views of quantum field theories, at the physical level, and we will see the details later on. And in conjunction with the all-enveloping relativeness at the physical level, the mathematical transcendence persuasively puts forward the very same scheme, manifested additionally at the level of mind. The justifiability of this argument emerges from the rather offbeat realization that our esthetical preference and creativity radiate the pitch of nonconvergence, of an identical magnitude. This may begin to move us toward piecing together all the elements—the subliminal as well—in the construction of an all-encompassing picture of the universe. Indeed, it would take many more inquiries and estimations before a holistic organization could be seen.

From the span of the observer that stretches alongside physicality, however, the emotive and the intuitive would appear as a part of the whole physical framework.

From where we stand right now, what appears to be rather flaky and less defined is nonetheless a forceful fact: the intuitive belongings amicably align with the exact same nonconvergence that space-time unabashedly packs. And that there is the one single stitch of mathematical order further draws our attention to the scenario that the gears of the subconscious too exist in relativity, through themselves and by the weave of space-time.

The two things are there because they coexist, in physiques, the whirs of nature that recite along those physiques, the spectacle of the entire universe that runs into the eye, the reflections that stir in the ground of the mental, or, again, the all-exclusive subliminal terrain against the full-length meadow of the cosmos.

Although there hangs about a slew of attestations—even excluding the bona fide affirmation that comes from the knowledge that seers communicate—that support the relative existence of the mental or the inner plane (and I will be getting to quite a few of these), I shall for now remain tied to the feature of transcendence.

The resonating transcendence in the universe that springs from the code of continual adjustments can be better envisioned after we have taken a plunge into the mystical boundlessness of π.

π

The Esoteric Pi (π): The Appearance of Curve

The Convergence of the Peerless Contour in the Dynamic Forces of Space-Time

The Configuration of the Curve

Can you measure a curve? Most of us will say, "Of course! Make it into a straight line and then just measure the straight line!" But the thing is, as soon as we convert it into a straight line, it is no longer a curve. Before addressing the deep-seated importance of categorizing the structure of a curve in reference to a straight toward geometrically assimilating the nature of space-time, we need first to take a cursory look at the influence that the seemingly plain curve–straight relationship has over the evolution of mathematics itself.

Humans have been drawn to savor the guise of the curve from even before the character of the curvature became a part of mathematical definition. The vivid inclination was apparent in the imaginative creativities from prehistoric times. Euclid of ancient Greece, who laid down the principles of geometry, utilized fastidious phraseologies, distinctively different from current usage, to address the phenomenon of curvature. For instance, in his renderings, he considered line as a type of curve and defined a straight line as a line that lies evenly with the points on itself, whereas he characterized a curve as a "length without width."[1]

Ancient geometers had employed many different articulations of angularities toward refining and augmenting the strength of mathematical commentary. As the evolution of mathematics continued, the clear seals of equations appeared toward formalizing the property of curvature, whereby the caliber of the curvature was seen in the window of mathematics, falling into specific niches. And it was by the maneuvering of curves that the compounded ambits of mathematics, such as topology, algebraic geometry, differential geometry, complex planes, and surfaces, prosperously transpired. And all of these phrasings bear a direct relevance to understanding the plaster of actual space-time and interpreting the observations that continually pop up in the field of quantum mechanics.

The integration of the peculiarity of curvature was preeminent all along the phrasing of the physical sciences, and the carriage of arching has been mobilized excessively in methodical applications, from ancient to modern times. The essence of the curve (as circle) was perhaps first deployed by Thales of Miletus (ca. 624–546 B.C.), a Greek philosopher and mathematician, in clasping a demonstrative theorem known as Thales' theorem. The curvature of the ellipse utilized by Edmond Halley (1656–1742) to describe the orbit of a comet, now known as Halley's Comet; the shape of a parabola utilized by Blaise Pascal in his geometrical lineaments and by Galileo in demonstrating the path of a projectile; and the use of a defined geometrical curve by Johannes Kepler in describing the laws of planetary motion are a few examples of the direct graphical adoption of the nature of the curve.

Then the logarithmic spiral, described by French mathematician René Descartes (after whom the adept mathematical efficacy "Cartesian coordinate" is named), plane algebraic curves to hyperbolic spaces, Riemannian surfaces, and, consequently, the coming of the manifold curvature are some of the advanced efficacies that garner the manner of curve, not only in furthering the authority of mathematics but also in deciphering the fundamental meanings of reality. The different announcements of curvature, and their mathematical makeup, can be relished online, under the Famous Curves Index from the University of St. Andrews website.[2]

Even more relevant is the fact that with relatively few exceptions of mathematical pronunciations, such as very basic articulations of arithmetic, number theory, and set theory, the propositions of mathematics all the way artfully tame the relationship of the curve with that of the straight line—directly or indirectly. From the elementary branches of plane geometry, algebra, trigonometry, and calculus to complex divisions such as the subdivisions of geometry beyond the Euclidean, advanced forms of algebra and its topological formulations that help weave the fibrils of quantum mechanical observations into a single harmonious tapestry, and the avant-garde facets of complex geometry that are essential to flawlessly modeling the sheet of space-time the way quantum mechanics telecasts, all the demarcations inherently carry, in one way or another, the rational of curvature against straightness—albeit masked in different ways, in addressing diverse dealings.

For instance, with the statements of calculus, we determine the rate of change, for a given process, by putting a tangent against a point on a curve (the function), in the case of differential calculus, or by assigning the area under the function, in the case of integral calculus. In both procedures we are looking for a closest possible approximation in relation to the delivery of the curve.

The Inherency of the curve in Mathematics and the Universe

The mathematical impression of "curve-versus-straight" becomes stringently thorough in leading mathematical descriptors—the scaffolds on which the current formulations in astronomical and quantum mechanical studies ensue. The eloquent manifold that the averments of Riemannian geometry embraces is an example of a sharply progressive subsumption of the furls of curve to the level of multidimensional scaffolding—a type of mold that has a preeminent role in concocting the modus operandi of space-time. For instance, complex partial differential equations that go into engineering the higher-dimensional topological structures simulate the continuum of the higher dimensions by the sequential acclimatization of the straight along the folds of the curve. In the later sections of this chapter, we will be able to sense that the character of curve swings only in relation to the mark of straight, both in the proverb of mathematics and at the level of matter.

One such superlative topological manifold is that of Calabi-Yau space, named after an Italian American mathematician, E. Calabi and a Chinese-born American mathematician, S. T. Yau. The Calabi-Yau manifold, which was contrived by German mathematician Erich Kähler (1906–2000)—who had formulated complex manifolds utilizing algebraic geometry percepts—models the extra extensions of space-time, the way it is conceived by physicists based on current understanding. The tools of algebraic geometry and differential approaches of calculus are employed in devising such illuminating graphical motifs. Here it is vital to acknowledge that the coming about of the complexions, like that of the Calabi-Yau manifold, was entirely a part of the mathematical growth and advancement, announcing the advent of a complex, circuitous, yet smooth topological establishment that justly integrated the prior understanding and became the basis for later developments. Their emergence as compelling utilities in the areas of physics was thus an entirely different ball game.

The six-dimensional furl of the Calabi-Yau manifold later became the key component in the equation of the string theory—the modern view of space-time that suggests that the structure of reality sways in eleven dimensions, where out of these eleven, six dimensions are proposed to stay heavily folded.[3] This is an extremely captivating assertion that we will return to in later sections, when drinking in the magical twirls of space-time and the voices that we hear from the camp of string theory. The mathematical statements of the Calabi-Yau manifold became the archetype in seeing the materialization of this six-dimensional bend.

All in all, curvature is the only way to the continuum because it is the only way to engineer smooth transitions that are free of sharp bends, creases, pinches, punctures, and catastrophic tears—whether in mathematical formulations or on the surface of the actual universe. As we will later see, this ratiocination is of immense value in interpreting the ultimate nature of the universe on the grounds of quantum mechanical tendencies and the potentialities that loom in contemporary judgments.

The naturally existing dispositions from quantum to cosmic scales, including the living systems, are contoured smoothly, bequeathed by the cincture of the curve. Hence all universal embodiments are a materialization of this "curve-by-straight" demeanor, because their spherical geometries emanate from the staunch relationship of circumference and diameter, titled as pi (π). Sinuosity is imbibed in the natural configuration, whether we speak of the entire universe or its constituting members: the disk of a galaxy, the silhouette of a planetary system, the bulge of a planet at the cosmic scale, and the imprints of molecules, atoms, subatomic particles, and their behavioral decorations at the quantum. The curve is the way by which the conscious anatomies are embodied as well, accounting for both the plant and animal kingdoms. Anything that exists naturally essentially is cast as a continuously bending line—the curve.

The alliance of "bend" and "lineal" equally permeates natural processes, such as biological growth and movement, and exponential decay. It is inherent to financial and economic trends. Anything that pulsates in the landscape of realism, even the intrinsic imageries that swell in the field of mind, essentially always involves a curved silhouette. The mental image of a quantum particle is never square or triangular; it is always spherical, even though no one has directly seen an imperceptible quantum entity. In the realm of imagination, any natural state, whether infinitesimal or gigantic, by definition will be nothing but orbicular; even the apprehended pictures of the central features, such as the duality of wave and particle, will appear leveled by the envelope of angularity.

The Perfect Pitch of Curve in Orchestrating the Concordance of Space-Time

In the light of the relativistic nature of space-time, the curve comes to be the brilliantly stroked configuration, pertinently blending into the geometry of absoluteness, where the ingredient of straightness would lead to catastrophic consequences. I previously identified how the spherical embodiment is the only mold that would fit into the relativistically dynamic scheme of space-time (see chapter 2, "Why the Happenstance of Three Dimensions?"), where the crease of straight would cause jerks in sweeping movements and bumps in the continuity of the fundamental forces, through which the movement is sustained. In fact, both movement and the forces are two reflections of the same order.

The attainment of spherical borders in the landscape of the ultimate appears to be a tactical materialization that takes place in compliance with the laws of the universe, where the deportment of roundness happens to be the essence behind the effortless flow that is necessary to uphold the optimal harmonic order of reality. For instance, the homogeneous navigation of the universal laws through all the cosmic bodies—which are spinning on their axes and revolving around a center—obligates that their molds be spherical. The force of gravity can act homogenously only when bodies are contoured spherically, for gravitational pull isn't a force that is exerted only between the two astronomical entities; rather, it is a field that continually interconnects every stellar object in the landscape of the entire universe. The earth whirls and wheels not just by the gravitational influence of the sun but also by the gravitational drag of all other solar system masses, and every other cosmological mass that gyrates and revolves in the same order. In such a layout, any steadfast progression under the influence of the all-encompassed gravitational sweep is possible only if the matter figurations are fashioned in spherical molds; any pinch or pointy edge in the conformation of certainty would evoke bumpy gravitational perturbations, concomitant to jerky cosmic motilities. Thus the setting in of the universal laws is conjoined with mass–energy forms being spherically contoured. Needless to say, the irregularity in the universal forces, and so the uneven mass–energy forms, are purely a hypothetical portrayal, as both the continuum as well as the symphonic undulations in the fabric of space-time rely on frictionless transitions under the umbrella of the unwrinkled forces.

A handle to grasp the tone of curvature is the argument of the mathematical pi—the most popular ratio in mathematics, denoted by the Greek letter π. The Greek mathematician and astronomer Archimedes was the first to approximate its mathematical value— to be greater than $3\frac{10}{71}$ and smaller than $3\frac{1}{7}$ (ca. 250 B.C.). This numeral expression parades nonconvergence, lacking any motif. Its boundlessness has been claimed to be measured for trillions of decimal digits:

π: 3.14159265358979323846 ...

The number π is a mystic mathematical expression that comes

to exist as soon as there is a circular built, which reflects a straight–curve alliance, ingenuously transcending from the known into the unknown. Where does it go? What makes it transcendental? Its ceaselessness is unfurled in the measurement that utilizes infinite series, engaging trigonometric functions; the mathematical functions of angularities are employed in the estimation. So, in the lexicon of mathematics, we have a way to comprehend the underpinnings of the straight–curled alloy.

The Machination of Curvature and its Message of Realism

The convergence of x/y would mean that x can be broken into y units. From the manner of π we know that the circle's circumference cannot split into its diameter units. Why is it so? The cutting of this Gordian knot can be helped by first diagnosing how the unit (or the smallest representation) of the circumference (the curve) compares with the unit of the diameter (the straight). The unit of the diameter would be a point in space. And the unit of the circumference would be a point in space as well. However, when establishing the ratio (i.e., when comparing the two in space), the unit of diameter will be a point, while in relation to that point the unit of circumference will be a curvature (figure 4.1). Regardless of how infinitesimal the smallest entirety of straightness is, the relative minimal totality of circularity will be nothing but a bend. Rephrasing it differently, the mannerism of the curve exists only in relativity (to the straight). The nonculminating attitude of pi exposes that the actualization of the curve in essence is a phenomenon of relativity; the actuality of spheroid cannot oscillate singularly. And the numerical string of the curve stretches endlessly in our thirst to calculate and assess.

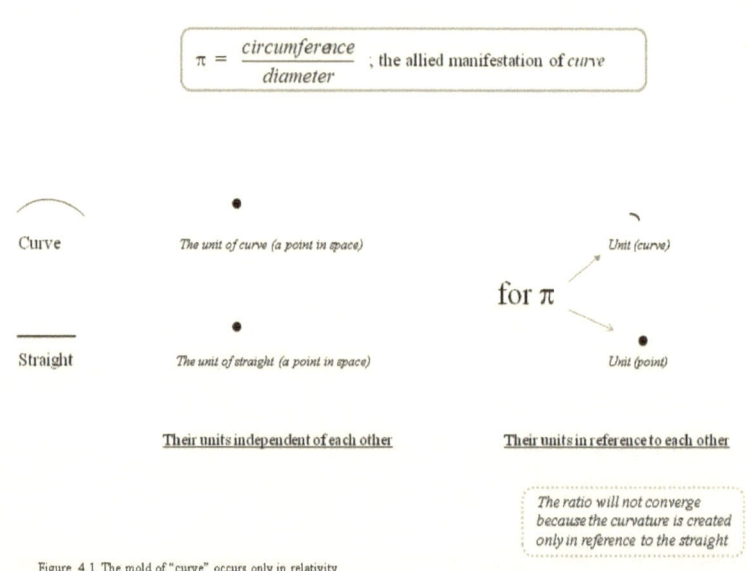

Figure 4.1. The mold of "curve" occurs only in relativity

Furthermore, the tenure of the tenacious relativity by the setup of the spherical forms will be implemented unconditionally, whether in relation to a spatial point within the sphere or from its outside. That is to say that the feature of nonconvergence seeps not just *into* all cosmic bodies but also *through* all of the cosmological makeup, establishing an all-around intercommunication. The overarching transcendence in the plot of realness echoes the skeleton of relativity in the constitution of matter. And we will see in the following sections that the transcendence of the ubiquitously occurring golden ratio (ϕ) also stems from this very curve–straight admixture.

The bearing of relativity in the complexion of space-time is a fact not new to scientists, or to anyone. As we noted earlier (chapter 1 under subheading "Procuring a Panoptic Window: The Picture from Physics") the "relativity" of reality is captured in myriad different ways; the interrelatedness of entities and the continuity of space-time both are forms of relativity. From Einstein's theories of special and general relativity, we know that the equivalence of mass and energy establishes the relativity between movement and time (a little out of context right now; I discuss this further in later chapters) and that the mass–energy forms exist relative to each other curved through the dominion of gravity.

Mathematical transcendence mirrors the existence of the same unyielding relativity and draws attention to the preeminence of curvature in synchronizing geometry to the characteristics of space-time. The numerical transcendence side by side also calls for soaking up the subliminal elements in the meadow of absolute collectiveness.

Part of the Same Game

Flagged by mathematics and advocated by theories in physics, the show of relativity means that the profile of earth lodges and goes on only as a part of the structural swirls of the solar system. The earth is there only because the sun is there (according to the quantum mechanical view, which I will discuss in detail later on). The architecture of the solar system subsists only as a constituent of the following systematization—the Milky Way galaxy; the actualization of the solar system is concomitant to the materialization of the larger gyration of the Milky Way, and so on.

As we voyage through the maze of mathematical proclamations and analytical deductions, I hope that you will be able to appreciate how the elements of conscious currents too flutter coordinately with the stirrings of the universe. The naturalization of aware eddies in the schematic of the natural laws isn't a matter of an abstruse philosophical viewpoint but is rather an analytical outlook based on mathematical textures, scientific notations, and conceptual feasibilities.

I relished the 2009 movie *Knowing*. Coated with paranormal psychic appeal and streaked with stunning visual effects, this apocalyptic science fiction flick features prophetic eventuality of cosmic scale. Although the plot does not point to scientific statements, the storyline nonetheless is greatly engaging, for, apart from its vividly mesmerizing setups, the anecdote reverberates a stylish blend of psychological dispositions and observed circumstances. In one scene, a professor (Nicholas Cage) discusses with his students the distance between the earth and the sun being the just right distance to set up a suitable set of conditions for life on earth to take place. On grounds of biology, one would say that life evolved on earth because its environment was physiologically conducive. Contrarily, from the commentaries of mathematical sciences, mainstream physics, and what we understand from psychological province, the pragmatic picture appears to be that of simultaneity—not just between the sun and the earth, but also in our encounter with them.

From the standpoint of the absolute watcher, to which the seers point, the spectacle of mass–energy and their operations en masse lie in a diagrammatic unison, and thus the life on earth; the earth itself; the sun, which makes the environment conducive to propagation; and everything else would be all stitched in the landscape of the unifying relativity.

Tying the Threads of Pi (π) and Phi (φ)

The ratios of pi (π) and phi (φ) emanate from the same mathematical scaffold. I noted earlier that the approximation of π was brought out by the acumen of Archimedes, and he made use of the structure of a regular polygon to approach this approximation. A polygon is a figuration that (1) can be inscribed by a circle (not all geometrical shapes can be inscribed by a circle) and (2) in many ways announces the fringe of the golden ratio in its configuration. Thus, at the graphical level, there lies a deep-seated one-on-one correlation between pi and phi. I will show shortly that it is again the pact of curvature to straightness that defines the transcendence in the thread of the golden ratio (ϕ) too. In the discourse of mathematics, it appears that genuine efforts to establish the direct equivalence between π and ϕ are made utilizing trigonometric functions.

We know that the golden ratio results when the ratio of a full length to its larger section is the same as the ratio of the larger section to the smaller one (figure 3.1). The cast of the golden ratio, however, symptomatically springs from the circular inscription hidden in the materialization of the golden ratio. The estimation of the golden ratio is made utilizing continued fractionation or by geometrical approaches, and we will acknowledge soon that the direct dependence of phi, the golden ratio, on the continuity of circularity at the level of straightness translates into the demeanor of recursiveness.

A line can be manually cleaved into the golden sections (the sections that associate by the golden ratio) by use of fairly simple procedures, and there is one available on the website of Ron Knot, of the University of Surrey. The procedure primarily involves making the line—the one we want to dichotomize—a part of a right-angled triangle (figure 4.2). The action of converting the line AB into the side of the right-angled triangle is the most decisive step because it is in this operation that the inscription of curvature takes form. According to Thales' theorem, if one side of a triangle is the diameter of a circle, then it is *only* the right-angled triangle that can be inscribed by that circle. Thus, in this correspondence, the diameter of the circle is also the hypotenuse of the triangle the circle inscribes, and the three vertices of the right-angled triangle are also the positions of the circle. In other words, the outlines of both the right-angled triangle and the circle are determined by the single set of those three coordinative loci (figure 4.2).

The rest of the method involves constructing two arcs: one with a compass centered at position C and stretched to position A, cutting onto line CB, and the other with a compass centered at position B and opened to the newly obtained point at CB, arching onto line AB. This new position, X_1, obtained on the line AB, divides AB into the golden sections; the ratios of AB to BX_1 and BX_1 to AX_1 exhibit the golden ratio.

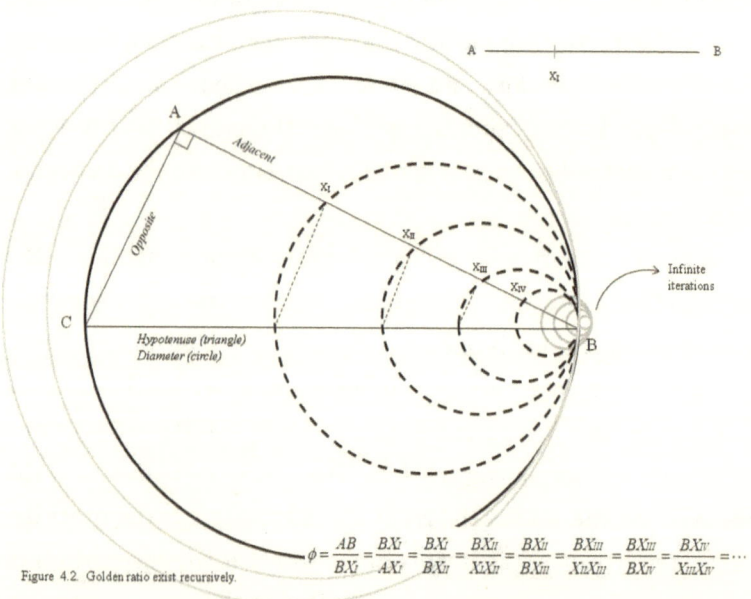

Figure 4.2. Golden ratio exist recursively.

The Relativity in π Is the Recursivity in φ

Underlying the cast of the golden ratio rests the phenomenon of infinite iterations of the lineament that led to the construction of the golden ratio in the first place. Thus the frame of the golden ratio displays in concomitance to the infinite proliferation of its own image in succession, through the launch of ceaseless iterations that sprout from the preliminary coordination, which engendered the golden ratio at the outset. Let us see how.

The line AB that splits into the golden section is the adjacent side of the right-angled triangle. The procedure briefed here is quite transparent but nonetheless requires arcs (curvature) to cook up the golden section. And although the appearance of the golden ratio in the argument of the right-angled triangle can be shown using the Pythagorean theorem—by way of the Kepler triangle, named after German astronomer and mathematician Johannes Kepler, who first saw the shine of the golden ratio in the triangle—the manifestation of the golden ratio is potently pillared on the involved recursivity structured by the shadow of the circularity in the build of the right-angled triangle itself (figure 4.2).

Hence, along with AB to BX_I and BX_I to AX_I, BX_I to BX_{II}, BX_{II} to X_IX_{II}, BX_{II} to BX_{III}, BX_{III} to $X_{II}X_{III}$, BX_{III} to BX_{IV}, BX_{IV} to $X_{III}X_{IV}$, and so on also are the materializations of the golden ratio that take place in concomitance (figure 4.2). The iterative constitution unavoidably manifests because BX_I doesn't form a golden section only with AX_I; it forms a golden section with BX_{II} as well. Similarly, BX_{II} doesn't just cut with X_IX_{II}; it also cleaves with BX_{III}. Every cut in this infinite series is in golden proportion with the subsequent slice as well as with the previous one. The reason behind this recursiveness is rooted in the cast of circularity that cryptically dictates the actualization of the golden ratio.

Resonating the ethereal tone of consonance, the appearance of the nonconvergence of the golden ratio is possible only if the sectioning is defined via the timbre of arc or in some way takes in the hem of a circle. The line AB not only echoes a circle but also golden splits through the shadow of a circle, concomitantly producing another right-angled triangle in the process. The resulting triangle again is an impression of a circle, and the adjacent side of this triangle again can be divided into the golden section, giving rise to yet another right-angled triangle and the associated circle, and the strings of φ burgeon endlessly. The chromaticity of φ iterates infinitely, and thus the golden cut always ferries an inexhaustible sequence of golden cuts (figure 4.2) in its temperament.

Besides being endowed with infinite gradation, a mathematical process that automatically repeats itself is associated with yet another structural axiom of central significance. At first sight, what just appears to be a linear array of circles and the associated triangle in effect points to the recurrence of the identical proportions—not just the golden proportion but all of the relationships that intercommunicate in the schematic of the curve–straight companionship: the circle in relation to the triangle, the sides of the triangle, and the relative position that cuts into the golden section all display the same structural correlation for every iteration. The key point is that whenever it comes to the self-sprouting infinite recursiveness, the everlastingness is a smaller part of the game. The underlying theme of intrinsic recursivity is the imprint of locked proportions between the constituents. And the dictum of the locked proportion is implanted by the bearing of the transcendental curve.

The ratio of circumference to diameter will always be the same, whether we refer to a Ping-Pong ball or the orb of earth. The force of π is universally invariant and is exerted irrespective of size. An analogous, somewhat trendy riddle is the rope-around-the-earth puzzle,[4] which challenges the comparison between the sizes of earth and a basketball, asking for the "extra length" of a rope that is required to encircle each item one foot off of its surface, compared to the length needed to band its surface snugly. Knowing that the circumference of the circle is $2\pi r$ (r being the radius), the extra length needed to circumscribe both the earth and the basketball will be 2π feet, or about 6.28 feet. The solution is actually free of the size difference between the two altogether; this is because the principle of curvature is a unanimous character perched on the pillar of transcendence and relativity, and therefore the extent of curve can be neither assessed nor differentiated. And owing to the immersed curve–straight relativity, the structural image of the golden ratio too essentially involves the procurement of infinite iterations, echoing size-irrespective measurelessness.

The exhibition of the golden ratio in the measure of the physique—discovered by German psychologist and mathematician Adolf Zeising (1810–76)—provides a straightforward glimpse into the setup of endless manifolding.[5] A section from head to fingertip is a golden section of the body's full length, because another section, from head to navel, aligns to shine the golden ratio in concomitance. The measure from head to navel itself again aligns with the measure from the tip of the head to pectoral to exhibit the golden ratio. The two fragments of the human body deflect the golden ratio only because there remains another fragment that consequently, and constitutionally, iterates to instigate a copy of the golden cut, with the preceding division (figure 3.2).

I have quoted only a couple of readings out of all the observed successive occurrences of the golden ratio in the human physique. The sketch of multitudinous gradation of the golden cuts can be found easily on the Internet, which can help envision the inborn recapitulations that come in the skeleton of the golden ratio.

The finespun weave of φ and π along the terminology of the curve—the conspicuous silhouette of nature, and the fact that the two numerical phrases flood the expressions from cosmic to quantum planes unequivocally—suggests the dwelling of a stately design that knits all the space-time elements collectively into one harmonious framework.

The Elements of World and Wit Sheathed in the Lineament of Curve

The diverse geometrical shapes that fittingly associate with the signature of the circle inherently also mirror the cast of the golden ratio, or vice versa. For instance, an equilateral triangle that can be utilized to generate the golden ratio also matches the enveloping of the circle on its vertices. The golden triangle, an isosceles triangle, ** too through the force of the curve can infinitely iterate to generate the chain of golden cuts, which in the incessant display forms the mold of the logarithmic spiral—the cast of a ceaseless arch, found in seashells, galaxies, weather patterns, fractals, and so on. Likewise, the fold of the pentagon, which can also be inscribed in a circle, also echoes the golden ratio (this was identified by making use of the work of Greco-Roman mathematician and astronomer Ptolemy [ca. A.D. 90–168]). The ratio of its diagonal to the side reflects the golden proportion because the diagonal at the same time also constructs the golden ratio with a pentagram that the pentagon successively inscribes (figure 4.3). Consequently, this pentagram

---

** a triangle with two sides of identical lengths, and consequently, two identical angles

shadows a pentagon in its center, which again displays the ridges of the golden cuts through its diagonals and sides, and sequentially through the diagonals and the pentagram that it inscribes, and the action goes on forever (figure 4.3). The pentagon and the accompanied pentagram iteratively self-sprout; and so do the pieces of the golden ratio. Titles such as *The Divine Proportion*, by H. E. Huntley, and *Geometry of Design*, by Kimberly Elam,[6] give a thorough account of the appearance of the golden ratio in the forms and angularities of plane geometry, including the triangular systems.

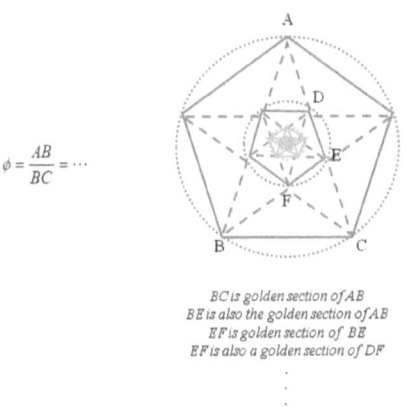

$$\phi = \frac{AB}{BC} = \cdots$$

BC is golden section of AB
BE is also the golden section of AB
EF is golden section of BE
EF is also a golden section of DF

Figure 4.3. Golden ratio materializes recursively. The pentagon-pentagram association that the circle inscribes iteratively display ceaseless chain of golden ratios.

All in all, the nature of transcendence, whether by π or by ϕ, is immersed in the ingenious geometry of the curve. Stated by the mathematics and stamped by the reality of the physical universe, the curve emerges to be the structural prime.

An easy, and intriguing, exercise can illustrate the primeness of the curve over the attribute of the straightness. We noted earlier that an equilateral triangle fittingly nests in the outline of the circle. Try making a freehand sketch of first the equilateral triangle and then the girdling circle that inscribes the triangle. The process of etching the circle along the vertices of the triangle not only feels tricky and awkward, but it appears to be almost impossible to pencil an offhand sketch to produce a reasonable-looking triangle–circle assembly. Now do the same exercise the other way around—making the boundary of the circle first and then drawing the tracing of the triangle within it. Not only does it feel a lot easier to construct the composition this way, but the resulting draft also exhibits a neater build and is visually suitable. I ran this experiment on an eight-year-old and an adult, without telling them beforehand what was expected. Both found it to be a lot easier if the circle was scribbled first and then the triangle was etched within it. And constructing the draft this way also secured geometrically commensurable pictures (figure 4.4). Curvature appears to be the defining factor. Evocatively, this experiment also sheds light on the conformity of the subconscious with the structural supremacy of the circle that enfolds the physical terrain.

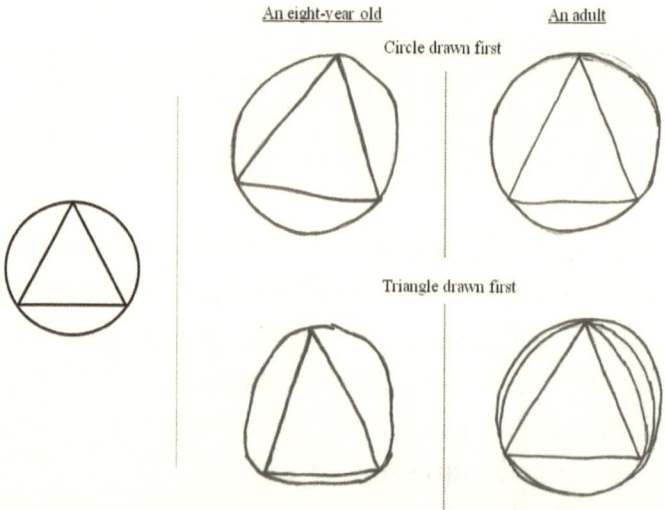

Figure 4.4. The primeness of circle. It is easier to draw a circle first, then inscribe a triangle inside it.

Squaring the circle is a perplexity that stems from ancient times. The challenge presented was to compose a square that has the area of a circle. With the authentication of π being of a transcendental nature, the task of squaring a circle was proven to be impossible; the aforementioned exclusivity of curve suggests that a circle and a square cannot be interconverted.

A square can be inscribed in a circle too. We can do the same exercise as previously for the assembly of a square and a circle. Again, it is nearly impossible to come up with a commendable sketch of the square snugly sitting in the circle, if we are to draw the square first. Tracing the circle first produces a seemly picture, not to mention that by following this route we fulfill the task almost effortlessly. Here again, the four vertices of the square symbolize the presence of the circle, and the diagonal line that cuts the square into two right-angled triangles also happens to be the diameter of the inscribing circle. If the two edges, the adjacent and the opposite, of the triangle measure 1 each, then according to the Pythagorean theorem, the diagonal line (or the triangle's hypotenuse) is formulated as

$$c^2 \text{ (hypotenuse)} = a^2 + b^2,$$

$$c = \sqrt{a^2 + b^2}$$

For $a = 1; b = 1,$

$$c = \sqrt{2}$$

Here we find another irrational number, $\sqrt{2}$, which can be utilized to announce the sequence of locked proportions via the ceaseless iterations.

The close rendering of √2, shows in the aspect ratio of A-size print paper. Cutting the sheet in half along the shorter side generates two sections carrying the same proportions as that of the original. Hence this action too innately carries the possibility of unlimited recursiveness. Allegorically and conceivably, the irrationality of √2, along with the unlimited stretch of geometrical repetitions, again disembarks via the fashion of the curvature. Expansion of more than a million decimal digits has been achieved toward valuing the irrationality of √2. The following sequence shows how it commences:

√2 = 41421356237309504880168872420969807856967187 5376 9480731766797 3799073247846210703885038753432764 1572 7350138462309122970249248360558507372126441214 97099 9358314132226659275055927557999 ...

Keeping in mind the geometrical kinship that plays between the inflexibility of the square and the smoothness of the circle, it is further intriguing to note that for any square figure (where all sides are of equal length; $a = b$), the area by the diagonal c ($c^2 = a^2 + b^2$) will always be an irrational expression. For instance, for a square that has the edges measuring 2 each, the diagonal c, according to the Pythagorean theorem, will be √8. For a square that has sides amounting to 3 each, the diagonal c will be √18, and for a square with edges amounting to 4 each, the diagonal c will be √32. Their irrationalities are

√8   2.828427124746190097603377448419396157139343 75075 38961463533594759814649 ...

√18   4.24264068711928514640506617262909423570901S6
261308442195300392139721974 ...

√32   5.65685424949238019520675489683879231427868753
015077922927067189519629299 ...

The area by the diagonal that results on summing the areas by
the two equal edges in a square will always be nonconvergent; the
invariable transcendental smear of the circle in defining this irrationality can be recognized, for, again, the mannerism of the square
is tossed by the continuance of the gyration. Yet another associated
masterly message is that the two squares of identical sizes cannot
agglutinate to yield a (larger) square.

Euler's number, e (2.71828182845904523536 ...)—named
after Swiss mathematician and physicist Leonhard Euler, who
broadly studied its properties—surfaces in various mathematical
descriptions and is utilized in various fields of advanced mathematics, engineering, and physics. Its transcendental nature was
proven by French mathematician Charles Hermite, in 1873, and its
application concerns the processes that draw in continual functionality, such as continuous compounding in finance or logarithmic
functions—pointing to the regularity of change by the adumbration of curve.

The aura of nonconvergence in the landscape of reality inseparably enmeshes with the ambit of the undying turn.

Decoding the Message of the Tangible Transcendence

The manner of mathematics is most commonly seen as a way to
quantify and correlate. The well-defined delivery of mathematics

paves the way to comprehending the intricacies of the real world. The illuminating voice of mathematics, however, is equally effective in the way of the nonquantifiable as well. The measures of π, φ and e cannot be quantified. Their lack of culmination under the influence of the curve, which molds every aspect of the physical and imaginary universes, unquestionably highlights a sweep of a harmonious accord across all that exists. The scenario of the lack of culmination is possible only if, as we have noted earlier, the "two entities" that we are comparing coexist and the manifestation of the one occurs in reference to the instance of the other.

The manifestation of the curve is the hymn of relativity in nature because from the interpretations on the behavior of π, we have recollected that the curve is created in concomitance to the measurement. And therefore it is not possible to appraise its actuality; anything that can be quantified in arithmetical terms will always have a linear, unvarying, constitution.

The ubiquitous nonconvergence in a diagnostic sense is a mathematical torch in illuminating the fact that every dot of totality cabins in respect to all other spatial specks, in the communal ambience of the materialistic plane, including the elements of wit. The observations from experimental and theoretical physics indicate the same modulation about the detectable arena, collectively—we will begin to gaze into it from the next chapter onward. Mathematical nonconvergence is an encryption of the same information in reference to our mental makeup too—the commonality of mathematical behavior between the mind and the universe accents the presence of a single overarching ordinance.

The glaze of the golden ratio in the expressions of the intelligence seems to be unambiguously definitive. The facets of intelligence flowingly formulate artistic objects, architectures, paintings, and symphonies that are intrinsically packed with the gems of the golden proportion. And it is not just in creative ventures, but in choices as well. The subconscious constitutionally longs for the transcendence of the very same mathematical scale that the physical universe innately possesses.

The all-embracing shadows of π and ϕ can be further seized for the plausible characterization of a missing link in the order of the cosmos—toward concatenating the current of the physical universe with the tide of the mind in the tapestry of space-time.

CHAPTER 5

Boundaries of the Unfettered Universe

A Cosmological Quest for the Misplaced Ingredients: The Secrecy of Antimatter

The Boundaries of Universe and Us

The colinearity of the mathematical transcendence between the universe before the eye and the one behind the eye also suggests that the conceived boundary between the mental and the physical tracts is, in the light of universal space-time, a sort of invoked fabrication. Pi (π) and phi (ϕ) as the embodiments of curve dwell from the cosmic to the atomic worlds, in various ways, and shimmer in our aesthetic preferences and pursuits, and imaginative depictions. The panorama of these mathematical patterns resides unfettered through the peripheries of both tactile and psychical extents. The authenticity of this categorical mathematical revelation that identifies the intuitive structure as belonging to the same field as that of the apparent universe must be accompanied by a justifiable explanation.

Conjointly, an analogous blatant query springs up as to why the universe is already structured in accordance with the way the psychic elements are attuned. In the light of the quantum mechanical observations, and their underlying synchronicity to the way the cosmos appears, there are two ways one can approach an understanding that can explain such an affirmative happening: first, that there is a kind of symmetry that is hitched between the latitude of world (external) and span of mind (internal), and second, that there isn't any actual "external" or "internal"—that all is one.

The insights from the physical sciences that can potentially endorse the colinearity of the tangible and mental planes emerge in two ways. First is through out-of-the-blue revelations that erupt in mathematical formulations of actuality. These numerical equations that crisply interweave the threads of analytical facts spewed novel findings, which later on were proven to be experimentally accurate, and they consequently gave a tighter foundation toward grasping the absolute nature of space-time. Second is by perceptively making sense of some of the stark differences of realism that emanate when we compare the analytical observations with the way the cosmos directly appears to us, to the naked eye or through telescope.

Before we go into the exploratory details, let us savor some of the sentiments on the reciprocity between the voice of mathematics and the contemplative sense of reality. Following are a few quotations uttered by some savvy mathematicians. These excerpts impart the poetical dexterity of these fellows, on the hunch about the parity between numerical codification and the idea of the utmost reality:

> The knowledge of which geometry aims is the knowledge of the eternal.
> —Plato (429–347 B.C.)

> Geometry is one and eternal shining in the mind of God. That share in it accorded to men is one of the reasons that man is the image of God.
> —Johannes Kepler (1571–1630)

> God created everything by number, weight and measure.
> —Isaac Newton (1643–1727)

> God does arithmetic.
> —Carl Friedrich Gauss (1777–1855)

> The infinite! No other question has ever moved so profoundly the spirit of man.
> —David Hilbert (1862–1943)

Self-evident…must not be confused with…
provability.

—Ernst Zermelo (1871–1953)

Mathematics, rightly viewed, possesses not only
truth, but supreme beauty—a beauty cold and aus-
tere, like that of sculpture, without appeal to any
part of our weaker nature, without the gorgeous
trappings of painting or music, yet sublimely pure,
and capable of a stern perfection such as only the
greatest art can show.

—Bertrand Russell (1872–1970)

God does not care about our mathematical difficul-
ties. He integrates empirically.

—Albert Einstein (1879–1955)

A flavor of the algorithmic cast of nature, and the attitudinal
bent of math adherents to fathom the order of reality, can be savored
in enriching chronological narratives such as *The Mathematical
Universe*, by William Dunham; *Why Beauty Is Truth*, by Ian
Stewart; *Unknown Quantity*, by John Derbyshire;[1] *Zero*, by Charles
Seife;[2] *Emblems of Mind*, by Edward Rothstein; and *The Language
of Mathematics*, by Keith Devin.[3] And we have noted earlier that
Max Tegmark, of MIT, has published many academic and general
articles asserting the ingenuity of mathematics in imparting the
way the universe presents itself, by his theory commonly known
as the mathematical universe hypothesis.[4]

So even if we disregard the power of mathematics in disclosing
the complex behavior of space-time or, for the time being, ignore
the sheer essentiality of mathematical expression in exposing the

composition of space-time, mathematics—in its light, ethereal portrayal alone—ostensibly murmurs the paint of the materialistic span. Let us now see the synchronicity of the hidden numerical messages with that which executes empirically.

The Coming About of Antimatter

The most stringent analytical testimony that can potentially construe the pounding of the inner plane symmetrically to the physical plane that meets the eye is the scientific realization of the occupancy of antimatter.

The existence of antimatter, which later was thoroughly and consistently verified, with remarkable accuracy, utilizing quantum mechanical procedures, was first discovered unintentionally by the pure revelatory power of mathematics. The possibility of the occurrence of antimatter was the marvelous insight of British theoretical physicist Paul Dirac (1902–84), in his attempts to come to terms with a causatum that loomed in his own mathematical formulation. Known as Dirac equation, this formulation concerns the relativistic implementation of wave. This mathematical expression, which was set out to blend the emerging quantum theory with the framework of special relativity—that is, space, time, and momentum as a single appearance, previously shown by Einstein—required interweaving of antimatter in the description of quantum fields, without which the equation ran into erroneous disorder and could not be rationally explained. The equation is as follows:

$$i\hbar \frac{\partial \psi}{\partial t} = (i\hbar \gamma_0 \gamma \bullet \nabla + \gamma_0 \mu)\psi$$

In the equation, i is the imaginary number (required in the characterization of the complex field—the mathematical field to simulate the quantum nature), is the reduced Planck's constant, Ψ is the wave function, t is time, and μ is the particle's rest mass. The expression "$(i\hbar\gamma_0\gamma \cdot \nabla + \gamma_0\mu)$" denotes an operator, with differential algebraic components, that incorporates special relativity into the quantum field.[5]

The preceding argument, a representation of Dirac equation, is an evolved form of Schrödinger's wave equation, which described quantum space-time prior to the assimilation of special relativity, and was originally proposed as a wave function of the electron. The notation of the relativistic demeanor of the quantum field in the preceding mathematical expression accompanied the need to incorporate the existence of antimatter because that was the only way to keep the mathematics of the complex fields accurate and flawless.

At the mathematical level, the correct account of the complex field, which describes the quantum field, means that positive and negative square roots must remain orderly and distinctive. The physical interpretation of this is that every particle must accompany an antiparticle that possesses a combative existence in reference to the original partner particle. This combativity, which was later experimentally verified, plays because although the particle and antiparticle are of the exact same mass, they are packed with opposite electrical charges or certain other internal quantum properties relating to particles' angular momentum and spin.

Ordinarily the square root of both positive and negative numbers is a positive value: the square root of 4 is 2; the square root of −4 is also 2. In a mathematically comprehended complex field—which recites the structure of the quantum terrain—the positive and negative square roots are treated separately, and this is achieved by incorporating an imaginary number i (see in the preceding equation), the square of which is set as -1: $i = \sqrt{-1}$. The exertion of the imaginary unit i is indispensable in the mounting of many advanced mathematical concepts, and its inclusion is vital in the equations that are evolved to describe the physical phenomenon of the universe.

The above equation is shown mainly for the sake of exposition, and many of the terms in it may appear highly technical for those readers not completely familiar with complex intricacies of mathematical physics. The myriad compounded details are not all that important here. Enough is the fact that the sightly gesture of this mathematical dialogue exposed the dwelling of a component, in the texture of space-time—the antimatter.

We would need to sink slightly into the picture of the quantum field and special relativity for weighing the significance of antimatter in the sustenance of matter, first, and then accessing the furthest view of reality, which would include the conscious dwell, by systematically sewing the shadow of antimatter to the cast of matter in reference to the observer.

The Upshot of Unifying Space and Time for Quantum Field

The terminological significance of field over particle in describing the quantum extent is connected to the way the spatiotemporal expanse would appear if we were to incorporate Einstein's special relativity into the quantum code. The theory of special relativity demands that a physical entity, of any measure, not present itself just as space; it must occur only as an amalgam of space and time. Thus the merging of the special relativity at the quantum level signifies an eventuality where a particle, as a definitive spatial entity, cannot hold, or else we would end up with an anomaly in the theoretical framework, whereby the space (the particle) exists aloof from the ticktock—the time. The existence of a particle both as space and time corresponds to a scenario where its discrete particulate form is not only generated but also, consequently, eradicated in the habitat of space-time: it's where the relevance of antimatter took the stage. (I surveyed earlier that the idea of antimatter, nonetheless, first sprung from the mathematical delivery of the quantum field.) For the uninterrupted flow of space-time, every quantum particle must accompany an antiparticle, and the ephemeral particle–antiparticle set, created by the borrowed energy from the field, must annihilate into each other, releasing the energy back to the rippling fabric of physical reality.

Paul Dirac's prediction on the dwelling of the positron, an antiparticle of electron, was experimentally confirmed in 1931. The rising of the positron brought the revolutionary inception of a fresh, and defter, space-time picture, expelled by the counteractive forces of matter and antimatter. The existence of antiparticle against all the fundamental constituents of matter is now a well-accepted, and proven, phenomenon, and several types of antiparticles have been consistently verified, on an experimental footing. A vast array of particle–antiparticle pairs, such as proton–antiproton, neutron–antineutron, muon–antimuon, and neutrino–antineutrino, are now known and well documented.

For those interested in glancing further at our encounter of antimatter in mathematical physics, the title *Antimatter*, by Franck Close, offers a light and thorough read.[6]

Mass, Energy, Movement, and Time: Different Pseudonyms of a Single Enterprise

Before we move further into the elemental flickering of the matter–antimatter pair toward interpreting the perceived vastitude of the absolute, it will help to first appreciate how the distinctive constituents of the universal makeup fall into a unifying web of space-time flow.

By the equation $E = mc^2$ (where E is energy; m, mass; c, the speed of light), in the theory of special relativity, Einstein avouched the interdependence of mass, energy, movement, and time. At first glance the equation primarily states that mass is energy, and vice versa; the element of c^2, from the basic understanding, came as a conversion factor. (Although it was Einstein who saw the set structure of physical world in this formulation, the interrelationship of mass, energy and speed of light was conceived earlier.[7]) However, magnificently, the interlocking of the factor c, which powerfully amalgamates energy and mass, in this equation further heralds the blending in of movement and time as well in the making of physical actuality.

The theory of special relativity brings forth the sublimity that the rapidity of movement determines the ticking of time, or how fast or slow a moment passes. The higher the speed, the slower will be the ticking, or the clock dilates (in reference to the observer of that speed). Indeed, such braided space-time effects become obvious only at a prodigious haste nearing the speed of light. Thus the doctrine reasons that the beat of equivalency doesn't drum just between the density of *mass* and the sparsity of *energy*; it throbs evenly between the play of *dynamism* and the catch of *time* as well. The interdependence of movement and time was a dexterous insight of Einstein that explained not just the continuum of space-time but its self-containment as well. The alliance of the extent of propulsion and the swell of the moment communicates that the feat of progression takes place within the framework of the unification, evoked by the fires of the universal forces. Let us see the mechanics behind such an edict.

In the full-length rhythm of space-time, the stunt of motion is not an arbitrary affair of an unconstrained displacement but is rather a progression that is tied between the two designated points; the travel always has a bearing for the beginning and the end. The electromagnetic radiation travels only if there already is a set destination for it to reach, or the electromagnetic force manifests only between the buckle of two objects—be it between quantum dots or cosmic globes. Likewise, the flight of light, the electromagnetic radiation that the eye can discern, sets off *only* because it is spreading to strike a termination point in the all-embracing granular vista of solidity. We can thus recognize the nimble layout of the movement–time plait, where within the framework of the structural dynamism, the passage of time would engage the same cruising, as light, that manifests between the two points of reference, within the entanglements of the universal forces. Hence, in the design of the equalized arrangement, the event of succession and the manifestation of time become two aspects of the very same potentiality; they are not mutually exclusive. For that reason, in the blanket of the constitutional agglutination, the click of time yields to compensate the instance of speeding. The higher the speed, the lower will be the frequency of the ticker, suggestive of the dilation of time itself. The dawdling of the clock will occur in reference to the observer of the speeding body.

In the breadth of the physical landscape, the bits of matter glide interconnected by the forces of the universe, especially by electromagnetism and gravity, where under the slick interlocking of time and movement, the speed with which light (the electromagnetic wave) travels becomes the maximal rate of motion that can possibly be achieved; this is the aphorism that sits underneath Einstein's explanation of special relativity. Because at an utmost scale the manifestation of mobility is the same as the establishment of the electromagnetic field between two points, the speed of this very electromagnetic wave (the light) is the maximum rate of progression possible in the anatomy of space-time.

The gamut of deductive specifics that I record here may seem a little specialized to some of us. However, I intend only to capture just the right ingredients needed to create a milieu in which not only the scope of the ultimate self-sprouts but also the spread of the consciousness and the fringes that dwell in that compass become self-evident. And scribbled here are the minimal nuts and bolts that are instrumental in whipping up the mise-en-scène of the absolute makeup.

We still need to cut across a little more of the scientific observations in which the offbeat flavors of physicality actually offer explicit insights toward picturing the full-length reality. For those who desire to dig deeply into the topic of these fundamental interpretations, there are felicitous general-read titles available that cover the chronological advancements in the understanding of space-time by the voices of mathematics and physics, with comments on the efficacies and the pitfalls all along the way—*The Road to Reality*, by Roger Penrose, is one such title that encompasses the historical and currently emerging interpretations in comprehensive details.

The apropos scheme of the space-time meld, in retrospect, also echoes that time at the junction of the observer also moves onward intermingled with the mobility, and thus, contrariwise, the slower motion, relative to the speeding body, will be seen by the shrinkage of time. That is, the moment will pass hastily in the ploy where every ingredient of the ultimate pool only oscillates by the fashion of relativity. The twine of the interlaced displacement and duration itself stays sustained by the shell of relativity (the more rapid locomotion occurs only in reference to the sluggish one), and so the variance in the ticking of the clock is again an eventuality that takes place solely in reference to the two different locations, in the animation of the space-time matrix.

A foundational feature of the macrocosm that is coupled to the theory of special relativity is the onset of ceaseless progression; the dwelling of all matter forms is chaperoned with the feature of their constant journeying, relativistically, in concordance with the ever-changing mesh of space-time. The theory of special relativity does not just describe the relative existence of mass, energy, movement, and time; it also highlights that each of these features plays an essential role, at any given point in time, in the continuance of space-time materiality. The cast of the realism would crumble if any one of the ingredients were missing from the composite that is put together evenly by mass, energy, displacement, and time.

The Essentiality of Annihilation in the Pulsation of Reality

The declaration of relativity that springs from the uncomplicated $E = mc^2$ is accompanied by additional axiological ingredients. One of such impartation is the flicker of variation in the equivalence of mass and energy. Mass is equivalent to energy only when it is accompanied by the drive of the movement. And although energy is a coalesced mass and mobility, the one that moves is always the mass, be it a quantum particle or a cosmic body. In the space-time camaraderie, this variation betokens the setting of a ceaseless lineup of transitions in any form—a displacement, a thumping of time or an evinced alteration in any other readable way. The establishment of the permanent mode of variation insinuates that any two loci in space-time cannot be equivalent, as these loci are defined not just by the particle that travels but by all the interactions that it encounters as well. Locus A and locus B cannot be identical not only because they are defined by a different set of perturbations, set forth owing to the constant dynamism, but also because locus B is itself a progressed form of locus A, where locus B comes to pass as a *future* of locus A, imitating the passage of time (figure 5.1 a).

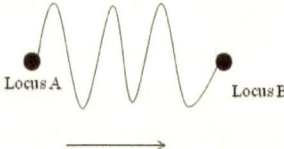

Figure 5.1 a. In the texture of space-time at an uttermost level, the stance of movement or the passage of time occurs between two loci.

Hence what appears to be the travel of a particle is more of a rapid succession of full-scale transformations in the synergistic setting; in the configuration of space-time, the voyaging signifies the setting up of the sequence of discrete sets of intercommunications, the orderly sweep of which would reflect the evolution of time. It is imperative to highlight that, in the geometrical sense of space-time, the vertical axis would be the reflection of space, while the horizontal axis would correspond to the echo of time—in the swaying of the all-inclusive blanket (figure 5.1 b). In context with the theory of special relativity, and the quantum mechanical phrasings, this vertical axis, denoting the measure of mass, would be equivalent to the horizontal axis, the measure of energy, once the imposed kinetics of the evinced realism has been accounted for.

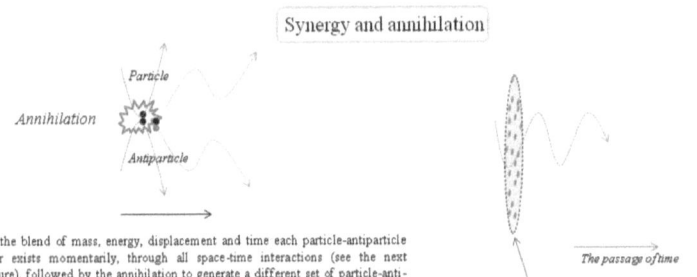

Synergy and annihilation

Annihilation

Particle

Antiparticle

In the blend of mass, energy, displacement and time each particle-antiparticle pair exists momentarily, through all space-time interactions (see the next figure), followed by the annihilation to generate a different set of particle-anti-particle pair, with a deflected path.

The passage of time

A time-point: Inter-communication and interdependence between all manifested particles creating a complex whole, which defines a time-point in the fabric of reality.

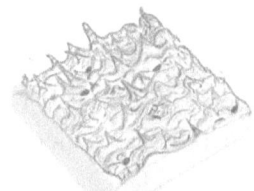

The space-time fabric. In the brew of mass, energy, movement and time every flutter denotes an embodiment and its subsequent eradication.

Figure 5.1 b.

It is the synergistic interplay—performed at the vertical axis (figure 5.1 b) under the brio of the key forces—that requires discretization, or the manifestation of the particulate nature. Let us now explore why the particle must annihilate to maintain the camaraderie in the lamina of space-time. In the orchestrated design of integrated physical realism, the progress of time implies transformation of one concatenated exposition into another, creating the two junctures of time. The crystallization of this exposition takes place by the order of the intercommunication and interdependence enacted in between every bustling shred that appears at that instant in time, adding up to form a complex whole (figure 5.1 b). Every instance of time transmits a picture of a unique integrated matrix in which each crumb of the concatenated ensemble carries through. The furtherance in the flow of time—or space-time—necessitates that every constituent materialistic crumb of the preceding time snapshot must transform into another matter-form in order for the following time snapshot to materialize. The process of annihilation compelled by the occupancy of the antiparticle makes the accord of this sheer transformation possible; the neutralization of those charges by which the particle–antiparticle pair holds its place in the structure of existence leads to the extinguishment of the particle–antiparticle combo-system, accompanied by the release of an unstructured energy back into the absolute framework.

Within the bed of the physical matrix the antimatter, possessing the same mass as the matter, carries a completely opposite charge and certain other quantum details, the pellucid feature by which the particle holds its place in the cadency of the interactive space-time. Thus the ephemeral appearance of the particle–antiparticle pair causes the eradication of both, concomitantly dispensing the locked energy. It is because of this annihilation that the energy across the entire field remains conserved, and yet the particles are persistently created, destroyed and crystallized again, and, foremost, the flow of time, or space-time, is maintained.

The manifestation of time, therefore, requires that matter, in any form, accompany antimatter, and if two particles can combine to generate a molecule in the existence, the corresponding antiparticles must also integrate to engender the antimolecule, executing the emergence of a mutually antagonistic pair.

Before moving on toward weaving the picture of totality by the highlight of antimatter, we need first to assimilate certain glitches that exist in our scientific interpretations, including the one with regard to the interpretation of the symmetry of antimatter.

Perspicuous Dissonances between Quantum and Cosmic Worlds

The realization of the built-in dwelling of antimatter in the quavering ocean of reality came from discoveries in quantum physics and, as we have seen earlier, the mathematics that goes into explaining them. The antimatter can be artificially produced in a sufficiently high-energy environment, and the fleeting presence of some of the antiparticles, such as the positron and the antineutrino, has been observed in natural processes. The pairing of matter and antimatter is now accepted as a fully verified aspect of our realism, and a large number of such pairings, some of which I have noted earlier, have been invariably observed.

Notwithstanding the undisputed inhabitance of antimatter and our acknowledgment of its scientific rationality in the operations of the physical realm, the cosmic universe almost entirely appears to lack this enigmatic entity, the antimatter. The cosmic realm does not exhibit the detectable annihilation would need to occur if the antimatter were part of the planetary arena. The straight-out absence of antimatter nevertheless attests to a paradoxical setting up of a colossal asymmetry in the wiring of the astronomic plane—in conflict with the quantum mechanical observation that declares the prevalence of sharp and determinate matter–antimatter symmetry, and at odds with the earlier made-out rationality of how this symmetry is an elemental feature in the flow of space-time.

The companionship of the span of pure observer and the dynamism of the physical makeup (detailed in chapter 1 under the subheadings "Procuring a Panoptic Window" and "Residence of Conscious Realm in the Trajectory of Space-Time," and elaborated upon in chapter 10) highlights a perspective that not only rationally tailors the scientific observations but also patches the contradictions that emerge from them. Needless to say, in such a fully overarching layout we would be able to see how the ticking of the conscious presence participates in the flow of the universe.

But even before reasoning for the stretch of observer toward dissolving apparent conundrums, it may help if we can first generalize why the discrepancies between the experimental quantum deductions and the firsthand appearance of the cosmos come to exist.

The Hidden Synchronicity between the Quantum and Cosmic Worlds

The heuristic formulations of space-time are mostly the concoction of what experiments show unfailingly and the way the universe appears to us directly. Classical mechanics, which include works of Aristotle, Archimedes, Leonardo da Vinci, and the paramount tower of Newtonian mechanics, deals with the fundamental principles of operations at the macroscopic level, from small mechanical objects to astronomical bodies. Quantum mechanics is the same kind of study projected microscopically—that is, involving the objects of atomic- and subatomic-scale lengths: the order that cannot be directly perceived with the naked eye. Conditionally tucked behind the cloud of our fanatical urge to understand the nature of reality, the mode of detection, whether in direct observation or indirect examination, may just be the sheer cause of certain disagreements—like the no-show of antimatter in the astronomical region I noted above—that are encountered between the point-blank of the cosmic scope and the roundabout of the quantum outlook. And taking into account this very mode of detection may further help us infer the synchronicity of our own placement in the matrix of space-time. Ergo, we might be able to spell out how the estate of consciousness itself couples along the paradigm that configures in the belts of the scientific explorations.

The mathematical predications that clump together the entire assortment of factual observations into a methodical whole provide a comprehensibly stringent and well-adjusted picture of the physical certainty. One such algorithmic formulation of the categorical unification currently peaking is the machination of the string theory, and the fitter models that derive from its nitty-gritty. The most up to date of these is the plot of M-theory. For readers who are interested in the technicalities of this silky doctrine, there are ample write-ups available, both in popular science and professionally specialized formats, communicated by mainstream scientists directly or indirectly linked to this field, such as Brian Greene at Columbia University, Michio Kaku at New York University, Leonard Susskind at Stanford University, and the well-received Stephen Hawking at Cambridge University. The purpose of bringing the scheme of string theory here in context of resolving the disparities between the quantum and cosmic fields is that this sleek mathematical recital of space-time, which incorporates all the universal forces and the observed elements, elegantly parallels exactly the way the quantum world behaves; the unification exhibited by the strings includes those features that we have missed at the cosmic level, such as the aforementioned matter–antimatter symmetry and the constant undulations, again a feature that is aired in the quantum mode.

Keep in mind that the cosmic and quantum fields are not two independent sets of operations; they are just two different ways of looking at one and the same corporeal arena. The traits of the antimatter occupancy and unremitting undulations, therefore, are part of the cosmological landscape, albeit they remain obscured, as much as they are part of the quantum one. The embracement of these ingredients along the system of the planetary plane must, in some way, be explainable if we are to see the equivalency of the basic principles in all the furls of the universe. In the surroundings of this vividly indicative unevenness, let us see how the "alikeness" between microscopic and macroscopic signatures can be reasoned methodically.

The optical sensing of an object is an advent of electromagnetic establishment between the object seen and the eye. Whether we notice day-to-day items, a cosmic body, or a far-distant galaxy telescopically, the instance of viewing jells because of the actualization of the electromagnetic radiation, transmitted by the object, striking the eye. Compared to experiencing quotidian articles and cosmic dots face-to-face, the quantum mechanical acquaintance comes through indirect surveillance. That is, we do not *directly* (one-to-one) run into the distinctive poses of the mass–energy spectacle; it is as though the palpations of the universe in apical harmony are eyeballed from outside its outer edges. Thus the contiguous (cosmic) and disjoined (quantum) worlds are two different portrayals of the same universe, one acknowledged from within the system and the other espied from outside the system.

Thus the quantum approaches proffer a full-blown behavioral picture of space-time and, in doing so, evidently display certain integrative factors that we cannot personally bump into when we eye the universe literally. This is because in beholding the universe naturally, the viewer becomes a constitutional piece of the full-dress plot. The window of quantum mechanics provides an uncut view of how the universe acts, plainly because this window is the only way to observe the working of mass, energy, time, and movement, without "we" becoming a constituent of the physical amalgam.

Reconciling Cosmic Antimatter by Accounting the Span of the Observer

The blatant lack of antimatter at the worldly level may point to the existence of a concealed plane, tucked away from the blunt connectivity that jells between the eye and the beheld. At the same time, the obscured extension must stand associatively against the worldly one, under the command of the overarching space-time continuum. The synchronous inhabitance of the inner breadth—in reference to the one that meets the eye—imprints a plane that is independent of the direct vista, and yet, in accordance with the enactment of physical laws, its harmonious blending with matter, movement, and time appears justly in effectuating the rippling of space-time—corollary to the significance of antimatter, which we confront at the quantum level, in the making of reality. This hazy-appearing argument can be seen as a conclusive dealing if we account for the fact of the plane of the observer extended alongside space-time. From the window of the observer the discrete display of the worldly plane manifests as a part of an all-sweeping spread of space-time. So in the view of the observer a symmetry beats, between the inner behind the eye and the outer interviewed by the eye, for every moment. This may seem a little overwhelming to absorb for some. I will elaborate on the symmetry of outer and inner in the final chapters.

In the reference frame of the matter–antimatter symmetry, it becomes pressing, however, to consider that the erection of symmetry would occur for every given point in time by the simultaneous rise of the subliminal and worldly drops, meeting to make a beat in the cadence of space-time. This argument again can be substantiated by the standpoint of pure observer, in relation to which a tick of the time alludes to the dissolution of the materialistic and inner plane for that instant—like the matter–antimatter annihilation we acknowledged in reference to the quantum flow. The participation of subliminal plane in the beat of time, then, would refer to surfacing of intellective and emotive elements. The quantum flow in reference to the observer, and the playing of the outer-inner symmetry, can be better appreciated in the context of the proposition of quantum history, which I will take up in chapter 9.

The whole sweep of space-time from the vantage point of the observer explains the implementation of definitive differences between the quantum and cosmic amplitudes. The tailoring of the accredited findings of the mass–energy equivalence; the wave–particle duality; the amalgamation of time and space, and time and movement; the quantum fluctuations; and the matter–antimatter symmetry in the description of the physical realness can all be simultaneously done just by accounting the clasp of the observer, and with it the articulation of the inner-plane-forged symmetry into the scene of the cosmos. I will schematically elaborate upon this later on.

This equilibrium of the inner constituents in the universal scheme doesn't just texturize the cosmological space into a striking whole; it also instills a formal equivalence between quantum and cosmic frames, for now, with respect to matter–antimatter pairing. Indeed, the portraiture of this view doesn't just stem out of a speculative guess; the idea of inner–outer coupling hails from tighter ground; that is, it is laid by the wealth of experimental findings and whispered in the latitude of the pure observer.

From the viewpoint of the pure observer, the playing of the matter–antimatter symmetry is uttered as an associative annulling of the internal and external (the world) frames, for every given point in time, in reference to the observer, which draws out aligned with (but independent of) the swirls of space-time. The discourse of the tangible sequence remains in equilibrium with the playing of the subliminal continuum, and from the perspective of the observer, the channels of mind and the grooves of what is caught by the eye remain the two aspects of the same configuration; there would be a single gestalt engulfing both the intuitive and the concrete layers. More on this will follow in the concluding chapters. I will now turn to how mathematics features the launch of the robust symmetry formalized between the elements of consciousness and the wiring of the visible sweep.

A Mathematical Glance at Matter–Antimatter Symmetry

Over and above the ticking of the symmetry between the out-of-sight inner (mental) and in-sight outer (tangible) is also vouched self-reliantly by the clean-cut tone of the numerical dialect. The recurring deluge of π and ϕ in every speck of the universe, along with the presence of the ditto π and ϕ in the creative aptitude and intellective volition, spotlights the same equilibrium being played between the actualization of the clairvoyant horizon and the materialization of the physical reach.

Thus the idea of interlinking the inhabitance of mathematical motifs evenly across the universal and phrenic domains with the occupancy of a designing creator can be refreshed to the occurrence of an inevitable simultaneity between the discernible universe and the extent along the mental ambits—a clutch of symmetry that self-propels in the unfolding of space-time. Both the ambits emerge and annihilate for every beat of time. And in such a scenario, the idea of linking the ubiquitous mathematical nature with the inhabitance of a supremacy that skillfully designed the exhaustive uniformity becomes obsolete. The transcendental demeanor of π and the associated ϕ, of the palpable and intuitive surfaces, translates into an overarching layout of homogeneity, which, when we relate to the occurrence of matter–antimatter pulsations with the recognition that the antimatter goes missing on the cosmological surface even though its mounting is decisively fundamental in the streaming of space-time, indicates the enforcement of the same underlying symmetry, mathematically presented between the show of nature and the subliminal.

The established equilibrium between the external of the sight and the internal of the subliminal (both segments flapping within the all-inclusive view of the pure observer), can be seen as an implementation that (1) is supported by the analytical observations and their theoretical frameworks, (2) cannot be refuted scientifically, and (3) is intuitively graspable, in light of how our own sensory faculties carry on. The description that the exposed and the hidden extensions of reality, reverberating the parallelism between matter and antimatter, being the two sides of the same materialization that makes an instance in time (figure 5.2), doesn't just accommodate all three of the aforementioned points but also untangles certain disparities that we encounter between the discerned quantum universe and the observed cosmic one.

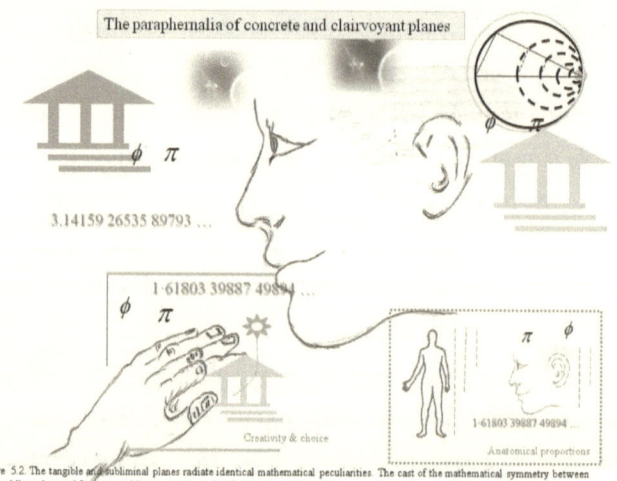

Figure 5.2. The tangible and subliminal planes radiate identical mathematical peculiarities. The cast of the mathematical symmetry between the worldly and mental frames would equate to the order of symmetry we discern from empirical outputs, which is vital to the flow of space-time

The well-documented accuracy of Einstein's mass–energy equivalence entrenched in the cosmic makeup—the equivalence that is not unambiguously obvious when we look into the macrocosm directly—can again be reasoned when we take into account the upshot of the onset of the sight in conceiving this equivalence. At any given point in time, the instance of seeing renders the noticing of the definitive aspect of matter, which, in connection to the movement, equates to the energy that reigns beyond/behind the sight. This ratiocination also has its bearings in the inferred indefinitive and abstract structure of the quantum universe, and in the panoramic vista from the standpoint of the undiluted observer—the topics we are going to come across as we go further.

The State of Grand Design

With regard to the perplexing context of the existence of a single dictatorial supremacy, mostly referred to as God, one thing that becomes unconditionally clear—from coupling scientific discoveries with obvious facts—is that all the elements of the physical and mental wings can be agreeably pieced together to accommodate all that we are aware of, and therefore the conduction of space-time is a trait that appears to be self-propelled in the continuum of the grand design. And in such a scenario, incorporating the "supreme governance" within the framework of space-time may lead to erroneous plights; space-time happens to be self-sufficient without it.

In the positioning of the traitless observer on the other side of the physical scheme, we can see the continuity between the inmost and the sensed apparency (of the materialistic plane) and acknowledge resolutions to some of the mysteries that we invariably detect at the quantum level, such as the whimsical waviness of the quantum realm against the clear steadiness of the cosmic territory. I will attempt to reconcile this in subsequent chapters.

The Setting of the Relativity between Beauty and Truth

Mathematics is a language of existence, and the universe and mind traces are the dosages of mathematics in one way or another. Mathematics is not just the nature of the universe; it is *the* universe. This idea was first formally presented by Max Tegmark in 2007. The functionalities of the subconscious sweep, which has the same numerical shadows as the outer, not only suggest the commonality between the two provinces that interknit to administer the loftiest spatiotemporal map but also indicate an aesthetical continuum that is jelled between the outer and the inner breadths. For that reason the phrenic constituents must be playing a definitive role in our sensing of the allurement in the appearance of the universe, where we are picking up the scent of beauty, not just because of the reflected brilliancy in the external corporeal truth but because the same splendor is echoed by the internal impalpable truth as well. On sentimental footing, it is plausibly because of this external–internal frequency alignment that the truth becomes beautiful. The book titled *Why Beauty Is Truth*, by Ian Stewart, is an exemplary account of the interrelationship between the sense of beauty and the truth of actuality that gleams in the mathematical tone of symmetry. For the conspicuous and the clairvoyant facets of the universe beaming the exact same mathematical phenomena, truth and beauty add up to just the two different ways of describing the same system.

The deeper we dive into the elaborate mathematical scene— which almost appears to be a self-guided labyrinth—the more deducible the universal disposition becomes, and math aficionados sense beauty in the exploration.

The Cosmic and Our Boundaries

The coalition of the inner plane into the cosmic landscape in reference to the observer is further stamped by arithmetic over the signatures of pi and phi.

Thus, when referring to an all-embraced continuum of reality, we can appreciate that beyond the instance of sight the boundary between the external and the internal does not stay put, or dissolve out, and all the existential ingredients coalesce as part of the same spatiotemporal scheme (figure 5.2). Our sensory perceptions in this scheme would play a key role in erecting an illusionary fence, thereupon segregating the pieces of the absoluteness of reality into the classifications of inner and outer.

This point of view has its origin along the decipherment relating to the skintight association of the observer in the coming about of space-time in the context of quantum findings, and in the positioning of this observer on the other side of the blatant reality in reference to the cognitively perceived reality. Both these frames of reference will sprout with specifics when we get into the topic of the nature of conscious existence and the meaning behind the manifestations of parallel universes in the hyperfield of the penultimate rhythm. (See chapter 9, on parallel universes.)

The mathematical commentary cogently resounds the dyed-in-the-wool truth of the all-encompassing unanimous scheme, where the reality of the visible and the inner tributaries enlace to mobilize a single overarching flow. Sketching the circle first and then inscribing a triangle or a square in it not only feels lot easier to accomplish, but the resulting outline renders a superior-quality image. The reason conceivably is rooted within the concerted design of the ultimate terrain—voiced in an arithmetical stroke—which blazes the primeness of the curve in outlining the plot of substantiality, both at the touchable as well as the subliminal levels. The settlement of the curve in the physical reality is mathematically synonymous to the effectuation of the transcendence, or of relativity, in all aspects of the all-inclusive framework.

In view of the relativistic demeanor of space-time that pops up in many different ways, such as in the teaming of matter with antimatter; the fusing of space with time; and the unwavering amalgamation of mass, energy, time, and movement, the succession of the intuitive ruffles too can be seen to materialize within the same associative framework—by the span of the observer. And in attempting to come to terms with the cosmological absence of antimatter and the cerebral impressions of the mathematical π and ϕ, the relative existence of mental ruffles is the standpoint that pressingly presents itself. The connected existence of the universe along the intuitive spread is a matter of deep emphasis toward understanding the continuum that not just sweeps blatant reality but also runs into the span of consciousness, where the concrete and the abstruse facets of physics transpire en masse. Perhaps that is the only way we can decipher the reigning continuum that is structured with the pronounced scientific findings that we witness.

CHAPTER 6

T

Witnessing the Boundaryless Structure of the Space-Time Landscape

The Compact and Diffused Fields Collectively Resonating with Quantum Flexibilities, alongside the Vigil of the Observer

The Span of the Universe: Straight and Stretched, or Warped and Compact?

Journeying into the mathematical realm is mystically phenomenal, often infused with a sense of clairvoyance. The more we delve into it, the more it captivates with truly bona fide projections of the macrocosm, in the most mysterious ways. This is why those astute navigators of the mathematical terrain that founded new branches, introduced important concepts, and set out new logics also emerged as distinguished philosophers. Often the natural zest for diving into mathematical ambits is accompanied by deep recognition for artistry—and the etherealness that emanates from it.

The stark image of reality in the plot of mathematics offers a way to access those deep-rooted intricacies of space-time that otherwise cannot be infiltrated. For instance, the reasoning behind the setting off of the endless expansion in mathematical texture could advisedly shed light on the coming about of the same behavior of the ceaseless spread in the expanse of the universe.

A related issue is rationalizing the straight, extended appearance in the display of the cosmic dimensions, which in the light of scientific arguments and mathematical expositions must remain curled and packed.

In light of the relativistic scheme, this eternal spreading out in the nature of the universe can be seen as the mediation of a rather deceptive perception that is effectuated by the exertion of the spatial relativity in the cosmic arena, which otherwise is furled and befittingly compacted, in all directions. Apart from accounting for the in-built relativity, this interpretation has its bearings in the empirical ascertainments at the quantum level and in the derived theoretical framework that emerges from them. I will discuss this a little later—let us first examine how the furling of space-time can be understood in the language of mathematics.

The orderly implementation of the space-warp, which is fundamental to the uniformity of physical principles, solicits that the embodiments of space-time must remain in a creaseless continuum, carpentering a state that imprints the boundarylessness in the conformation of reality. And for the play of this smooth continuum, free of fissures and sharp creases, the elements of the space not only must exist blended with the integrated whole but must also pass through each other, frictionlessly. At a quantum field level we see this as the continuation of manifestation and annihilation.

In the system of the well-defined and unambiguous universe that we eyeball, the frictionless passing of one form through another may seem a little of a bizarre conception, as that would imply that the perceived singletons in reality are diffused actualities, where in a quantum field rendering, each one of them spans the entire stratum of space-time. In such a mechanistic scenario, our own dwelling is more of a state of dispersive habitations, where beyond the five sensory perceptions, *we* remain diffused, spanning every nook and cranny in the field of reality. And at any given point in time, *we* lie traversing all the forms and sights that we come across. Before discussing the possible meaning behind such a whimsical nature of our habitation, let me illustrate what the exhaustive borderlessness possibly augurs, by an example of a simple mathematical ploy.

The Möbius Transformation

The formalization of multidimensional space first took hold in the reaches of mathematical narration—soon following the advent of the system of complex coordinates—via the works of Irish physicist and mathematician William Rowan Hamilton (1805–65). From the extension of the two-dimensional complex coordinate, Hamilton contrived four-dimensional articulation, known as quaternion, bestowing a rich leap not just in the progression of the algebraic voice but also toward grasping the curved (relative) nature of the real space. The adaptation of further algebraic intricacy took tenure soon thereafter. The one that we intend to relish here is the modern algebraic machination of the Möbius transformation, which germinates by the script of homogenous coordinates set out by German mathematician August Ferdinand Möbius (1790–1868), who utilized the previously ascertained four-dimensional mathematical structuring.

Möbius transformation is a mathematical procedure of conformal mapping that leads to translation, dilatation, rotation, and inversion of the complex plane. A detailed depiction of this transfiguration can be seen online, on the webpage offered by Douglas Arnold and Jonathan Rogness of the School of Mathematics, University of Minnesota.[1] Translation specifies moving an object such that the relative distances between the constituent elements do not change. Dilatation relates to uniform expansion or contraction, and rotation to rigid body movement around a fixed point. In reference to envisaging the recitation of the boundaryless spread in the mold of the palpable reality, among all types of transformations the inversion is of particular enthrallment. By applying the process of inversion, it is possible to turn a geometrical sphere inside out without creasing or rupturing its material. The underlying feature of such a potentiality is the unhindered passage of the sphere's material through itself. The functioning of this mathematical elasticity is revelatory of the mechanism by which the indispensable malleability and permeability that permeate every component of space-time beget the boundaryless continuum. The universe that we apparently behold as stably solid at its core abides fluently disseminated, whether we reason based on cues from quantum physics or rationalize based on the structuring role of sensory faculties in our perception of the universe as impenetrable and inelastic; the matter at its kernel remains pervious and indefinite. The abstractive nature of the wave, in concomitance with the existence of discrete particles; dispersive energy against the coexisting solidity of matter and the fluidity of apperceptive terrain in comparison to the compacted mat of the observed universe that we meet up with, all stamp the seep of elasticity in every capacity of the grander reality. The coming about of the Möbius transformation in the province of mathematics brings forth a portrayal that can imitate the flexing in the girdle of the unifying continuum, leveled by the cast of boundarylessness.

The process of Möbius transformation involves isomorphic (*isos-*, equal; *morphe*, shape) transformation of the object—a complex plane, say, x, to its inverse function, x^{-1}, such that x and x^{-1} stay homomorphic (*homos-*, same). Such topographical transformation is also known as bijection—precisely because this progression involves one-to-one mapping between the two sets of elements such that no unmapped element exists. In a broad sense, the process is corollary to deconstructing an object into small dots, rearranging those dots with respect to each other, and reconstituting those dots into an object that has identical shape and aspect ratio as that of the original.

The transaction of Möbius transformation relies on a fancy-looking rational function:

$$f(\zeta) = \frac{a\zeta + b}{c\zeta + d}$$

In this equation, a, b, c, and d are the complex numbers denoting numerical impressions in a complex matrix, meeting the condition of $ad - bc \neq 0$, ensuring that the matrix is invertible; ζ is a complex trigonometric function, for the transformation.

As they appeared quite a few times earlier in the text, it is relevant to put down a few words on the voicing of the complex numbers. Complex numbers are the numbers that have real and imaginary parts, where the imaginary part is suffused with an imaginary unit i (which we noted previously in the section "The Coming About of the Antimatter" in chapter 5), and the complex plane is a geometrical representation of the complex numbers constituted by a real axis and its orthogonal imaginary axis. The symbolization of the complex plane became a crucial tool in the astute description of intricate functionalities, such as in electromagnetism, hydrodynamics, electrical engineering, and refined mathematical fields, as well as, as discussed in the previous chapter, when comprehending the elaborate nature of the quantum terrain. The imaginary unit was used in finding solution to algebraic equations or, as noted in the preceding chapter, in handling situations on a Cartesian plane where a square of a negative number is required to be a negative value only. Cautiously avoiding going too much into detail, let's just say that this imaginary unit i, a part of complex number, allows numerical solutions for every element in the geometrical plane. Let us now ruminate on how the spatial elasticity can be grasped first by the telecast of physics and next by the stamp of mathematics.

Space-Time Elasticity and the Integrated Picture from the String Theory

The significance of Möbius transformation in pinning down the actual penetrability and permeability in the temper of space-time can be sensed better after a glimpse at how scientific research advocates the dwelling of flexibility in every crumb of the operative universe—the universe that we perceive as a panorama of compacted bodies. From the intellections of wave–particle duality (that is, the wave and the particle being two aspects of the same reality), Heisenberg's uncertainty principle (that is, the inability to determine simultaneously the position and momentum with equal precision), and Einstein's mass–energy equivalence, to the up-to-date interpretations of the physical principles, the methodic explorations affirm that the opaque and impervious universe that we come across indubitably operates with penetrable elasticity. And in the light of the current ratiocinations, the surfacing of the string theory effectively brings forth a consentient scheme of harmonious malleability, scripted through each and every constituent of the factual arrangement. Consequently, as noted in the succeeding sections, the passable flexibility is also executed in the setting up of the dimensions, establishing a wrinkleless continuum—a state of boundarylessness—in the dynamics of the space and time.

Since its first revolutionary inception, in the years from 1984 to 1989, the string theory has been widely relished and academically savored for its striking proposal to establish dynamic consonance between all the known elementary particles and the fundamental forces of the universe. (Reference articles: Green and Schwarz, 1984; Candelas, Horowitz, Strominger and Witten, 1985.[2,3]) And on the grounds of advancing mathematical physics, the string theory has been the one and only architectural paradigm by which we may be able to substantiate a holistic view in which blobs of matter interactively pulsate under the influence of the coalesced forces in the harmonious pool of space-time. Often touted as a theory of everything, the string theory certainly does extend a full-blown picture of a dynamic uniformity. *The Elegant Universe*, by Brian Greene, presents a thorough and popular science–format account that covers the background and advances of the string theory, whereas the written narrative *Superstring Theory I and II*, by Michael Green, John Schwartz, and Edward Witten, is a text-book-format account that covers details with greater mathematical depth.[4]

According to the string theory, nature at its most basic level dwells in infinitesimal single-dimensional strands of vibration—labeled as strings—of a scale around the Planck length (1.616199×10^{-35} meters). The string's frequency of vibration, an expression that stems through the interactive force field in the bed of space-time, imprints the peculiarity of a mass particle or a force particle. And thus, in the model of the string, all types of fundamental particles as well as the universal forces are ascertained just by the distinctive modes in which the string vibrates. Just like the way the formulation of special relativity into the quantum canvas gave us fresh insights, leading to the discovery of antimatter, the mathematical genesis of string unveils certain atypical characteristics that on syllogistic grounds appear to furnish a sound blueprint of the way the universe is.

Incorporating Einstein's general relativity (that is, the gravitational force) into the string formularization in the form of the graviton—a force particle for gravity—provides a framework of a communal space-time, with utter precision, where the fluctuating distinctiveness comes by only as various modes of vibration in an otherwise unanimously transformative, all-encompassing back-ground. And although there has been a significant deal of skepticism concerning the scientific validity of this hypothesis in displaying realness, both at the general public level and by the mainstream members of the academic community, so far the schema of "string" has been the only way to see all the physical phenomena truly assembled into a coordinative whole.

In the game of this orchestral wavering, thus, the specific frequencies at which the string oscillates don't just connote the embodiment of matter but also insinuate the incarnation of the four fundamental forces. The embracement of the universal forces in the forms of the gluon and weak gauge boson (the force particles for the strong and weak nuclear forces, respectively) and photon (the force particle for electromagnetic force) rests on the theoretically recognized and experimentally established fact that declares the particulate inhabitance of the aforementioned universal forces, accordantly authenticating the resonating string model. Whereas the embodiment of gravity, designated as the graviton, has yet to be seen empirically, its occurrence, nonetheless, is widely accepted on ratiocinative and deductive grounds.

Therefore, in the schema of string, the landscape of coordinative vibrations is studded with the matter as well as the forces by which the matter itself subsists. Hence, in retrospect, the string theory treats the panoply of matter and the implant of the forces as two sides of the very same coin. This archetype of reality thus allows a wall-to-wall unanimously alterable landscape for every given point in time, or space, precisely because every aspect of the physical nature is now distinguished only by the mannerism in which the most intrinsic filament resonates. In this aqueous framework, which is attuned to matter's formation as well as to its disintegration (matter–antimatter annihilation) coequally, every nook and cranny stays infused with a dab of elasticity—a feature that is acute for frictionless flow in the boundaryless continuum.

Analogous to the discovery of antimatter by the mathematical statement that was set out by Paul Dirac, the coming about of the superstring surmise—currently the same as the string approach, which attempts to meld matter and all of the forces into a single presence—points out certain peerless features that, if within reason decoded equitably, can not only extend a sharper and clearer picture of the absolute reality but can also shed light on how our own presence is established in this irresolute landscape of continual fluttering. And although some internationally acclaimed physicists debunk certain judgments that spurted from the string configuration (I have enjoyed stark apprehensions raised by Roger Penrose—not to mention his gleeful tone on this subject, which I found quite amusing—a physicist and mathematician at the University of Oxford, in his national best seller *The Road to Reality*), it surely does deliver a promising picture of the way the universe befalls, where not only the scientific validations reside but also a description of our own location in this majestic weave dwells; I will come back to this. For now, we need to keep up with the disclosures of string theory in order to picture the operation of the boundaries.

The Quandary of Dimension

The second superstring revolution brought further genuineness to the string schema when Edward Witten, a theoretical physicist at Princeton University, assembled five types of string theories, which had stood independently until that time, into an overarching eleven-dimensional postulate, which he titled M-theory, by showing how these five different forms of arguments are pillared on a single architectural framework, known as duality.[5] This bright insight gave much firmer ground to the conceptualizations of the stringlike phenomenon. However, even after drastic advancements, some of the looming specifications still caused many leading fellows much leeriness.

A dubiety that has evoked compelling concerns is the issue of the number of dimensions that the physical arena reaps in the string mold-based discernments. The mathematical formulation of this physical scheme under the paradigm of vibrating strings takes form only if space-time is conformed in ten dimensions instead of the regular four (three space and one time) with which we are familiar. Despite having faced clear unrest in the scientific community, the root cause of the materialization of these ten dimensions has not been clearly conceived, either on mathematical grounds or by the linkage of such a scenario to the real world. The most prevailing explanation, according to the mainstream string theorists, is that out of the nine spatial dimensions, six dimensions remain curled on an infinitesimal scale such that we are not able to see or perceive them, while the rest of the three spatial dimensions underwent unfurling and lodged in extended configuration as the universe formed and continued. Such an explanation, although it can be force-fitted into the framework of the string phenomenon, is very weakly justifiable on scientific grounds and is not intuitively appreciable.

The biggest concern that bobs up is why six of the nine spatial dimensions continue to submit to compaction, whereas the remaining three nests expanded. Our scientific and mathematical course of thought, on top of instinctive sense, would dispute such an inequitable falling out of the dimensions—an outcome that in all likelihood should be naturally driven in accordance with the uniformity of the universal laws. And under the edict of the overall evenness, all the dimensions must remain on equal footing and must collectively beget a justifiably symmetrical manifold; that is, all the dimensions must remain congruent to the extent with which they are curled up or extended, even if we were to consider the real universe to be swaying in ten dimensions, instead of in the regular four that we directly notice.

Quantifying the Dimension of a Terrain That Is Perfectly Pliable

The flexibility-imbibed indistinctness is a trait that is essential for the smooth flow of space-time, whether we talk of its constituting elements or the erection of its dimensions. The institution of dimension is a natural outcome that occurs in accordance with the launch of the omnipresent forces, in much the same way that matter is geometrized optimally in conformity with the comprehensively consistent natural prescription. Thus the erection of dimension, or the number of dimensions, cannot be just a happenstance that is impervious to the way space-time befalls. The number of dimensions and their anatomy in conjunction with the arrangement ought to, in some way, correspond to the conduction of reality; and thus, in our ascertainments, space-time geometry as well as its number of dimensions must agglutinate within the same unified universal scheme that we are longing to discern.

The optimality of curve in establishing the synchronicity between matter, mobility, and the universal forces, as we made out in chapter 4, can be interpreted as equally pronounced in the manifestation of dimensions as well, in bracing the flawless streaming of space-time. Toward attempting to deduce the facet of dimensionality, we must first acknowledge that the number of dimensions that we account as being three in exactness do not stay stretched in three discretely defined vectorial directions but rather lie converged into each other via the inflection of the incalculable curve. And thus the natural ambits, individually as well as in totality, exist by the character of transcendence, whether we talk of entities, quantum to cosmic, or, by the same syllogism, the spatial dimensions by which we are trying to discern the topology of the ultimate turf.

The bewildering issue of the number of dimensions, or how they are set, might just be springing from the definition of dimension itself, which we have driven to formalize, where the outcome of the multitude of dimensions in the mathematical formulations of actuality might be due only to our accounting of the three spatial dimensions as three *distinctive* extensions in the first place. While an object such as a cube accurately fits in accordance with our definition of dimension, by the pitch of three independent amplitudes, in a sphere—a ubiquitous natural mold—on the contrary, the display of discrete dimensions remains ambiguous. And along the same indication, the dimensions of the universe that we perceive as open in three extended dimensions in actuality may just be transcendentally curved, all converging abstrusely into an indefinite sphere-like object. Therefore, whether we talk of a three-dimensional entity or the three dimensions of the spatial stretch, the magnitude of the prolongations in the sinuous field of the ultimate intermingling may only highlight the inducement of the three-way transcendence, that unanimously converges into a unitary conformity of a complex manifold.

The proposal that a certain quantity of spatial dimensions remains curled up in the imperceptible domains of space-time, on indicative and affirmative regards can be seen as trendy way-out explanation to tackle the overflow in the number of dimensions that surfaces in the mathematical model of the string theory. This explanation seems a little outlandish both at the scientific and intuitive levels. At the most basic level, the emergence of the multitude of dimensions in itself might reflect an artifact, structured as a result of feeding three distinct dimensions into the formulations of reality.

The Frame of Harmoniously Merged Dimensions: The Realer State

The spatial dimensions that we see as a mirage of straightness are topologically converged into a complex manifold. This interpretation stems from the realization that nothing in reality exists lineally. Assertive cues from the way the quantum world behaves, the understanding of the wave–particle duality, and the declarations from theories of relativity all pronouncedly point out that behind the guise of "discrete and lineal" lies the certainty of "disseminated and curvilinear." From the standpoint of the ultimate observer, the mark of "discrete and lineal" and the quirk of "permeable and incurvate" lie as two distinctive aspects of the same physical arena, one existing via sight (or the five senses) and the other from beyond it (them).

(The graphic of observer has frequently turned up in the past few chapters and here, and may seem a little without proper context. My bringing the reference of observer is an attempt to corroborate the accuracies of systematic interpretations through the spectacles of the observer, and the subjects of some of the posterior chapters are all but fully dedicated to the attribute of the observer that seeps into scientific interpretations and sentient presence.)

The intricate consolidation of the spatial dimensions can be neatly depicted by the previously noted mathematical structure of the Calabi-Yau manifold. The Calabi-Yau structure, which at first surfaced in the branches of algebraic geometry and topology, bears immense value in conceptualizing the existence of multidimensional fields, the order of which is of weighty significance in describing the universe by the beau ideal of the string theory. This topological fabric neatly maps the convergence of the multitude of dimensions into a unanimous wholesome singleton. And although in numerical terms such a morphological display announces the presence of the assemblage of dimensions, in the consideration of inseparability the form reflects only the feature of transcendence, studded at its every locale, and thus portrays dimensions that are neither individualized nor carry any significance as isolated peripheries.

Mathematics is not only about precision and quantitative transparency; it expresses the abstractive too, in most striking and vivid ways. And so, by the demeanors of mathematics, we may be able to decode the innermost features of truism, which otherwise, at the level of physical perceptivity, could remain inconceivable, even if all the experimental outcomes too culminate in pointing toward that very same design of reality that the algorithmic tone is exposing. Whether we talk about the transaction of Möbius transformation or the portrayal of oscillating dimensions by the entwining of the Calabi-Yau manifold, the realization of such intimations is possible only if the boundaries, which journey along the structural rims, are taken to be penetrable and fluid, weaving an unvarying mobility through every speck of the expression.

A transformation such as the inversion of a sphere demands the incorporation of a functionality that, although allow motility, instills containment. Known as "point at infinity," [6] the introduction of this mathematical facility on a number line results in a closed curve, where the two lineal ends abide continuous to each other.

The point at infinity interfused with the homogenous coordinates, which we noted earlier, constructs a projective plane, thus transforming a complex plane into a closed–surface complex shape—an algebraic sphere that exercises one such point at infinity. This geometrical sphere, called the Riemann sphere, named after German mathematician Bernhard Riemann (1826–66), blazons a point at infinity at its pole. Thus, in this projective plane, the function of a point at infinity administers an implicit boundary. The elements of the object, then, although restrained from diffusing from the margins of the structural domain, float fully unobstructed. Introducing a point at infinity warps the space along the fabric, and as a result, that fabrication emulates the modulation of a self-contained universe. Its elements cannot see what lies extramurally, for their interactions hang within the thresholds of the morphological framework.

A geometrical range comprising complex numbers is referred to as a complex plane, and when this surface is augmented with the point at infinity, the resulting regularity becomes an extended-complex plane. The statement of the extended-complex plane puts forth a picture about an adaptive orderliness, perfected beyond the territorial grapples of the peripheral restrictions. By using mathematical functions, it is straightforward to invert a sphere because its elements can permeate each other, and yet the shape of the sphere remains preserved by the punch of the space-warp, fashioned by the kink: the point at infinity. My objective of bringing in the subject of such an emblematic transformation and the functionalities that actualize them is to reiterate the truthfulness in mathematical manifestation by which we may be able to conceive the modus operandi of the reality that otherwise is not downright revelatory.

The pervious elasticity that we appreciate in mathematical profile is the underlying nature of the actual matter, as is cued in the voices of experimental measures and discerned through the glasses of the watcher that aligns not just with the solid view but also with the five sensory faculties of the supraliminal presence. The attribute of porous adjustability, in turn, is an essence that allows transformability—a quintessential factor in the flow of time, without which even space doesn't become evident.

A topological form of Möbius strip—independently dreamed up by German mathematicians Johann Listing and August Möbius—dispenses yet another specimen for the composition of the boundaryless continuance. Take a strip of paper, twist it around halfway, and glue the two ends together. This results in a ring with an embedded twist. Merely introducing the half-twist leads to the transformation of the strip's physical properties. The act of turning followed by cementing the two edges together produces a band that has double the length of the original strip, and, interestingly, the resulting band engages only a unitary surface. Cutting the transformed strip down the middle through its entire length will bring about a single and broader orbicular band instead of two smaller ones. The outcome of the twist in the structure of the Möbius strip, at the physical level, is analogous to the execution of boundlessness in the projective plane, where the constituting elements traverse through each other readily. The act of affixing the two ends is parallel to augmenting the point at infinity of the projective plane. The ploy of shaping a spiral halfway thus trickily dissolves the demarcation of the inner and outer, rendering the effectuation of a homogenous singular surface, lying beyond the impediments of fringes—for one set of edges, in the case of the Möbius strip.

The notion of the mind and the universe being two distinctive existences emerges in concomitance with the manifestation of the virtual boundary that parts the integrated "exhaustive" into the internal of mind and the external of universe via the execution of the five aware senses. And whether we account for the holistic view of space-time from the perspective of quantum mechanics or acknowledge the effectuated physical reality as being the universal sweep of the unitary total from the viewpoint of the watcher, the disposition of the boundaryless permeability turns up as an underlying character of the fundamental makeup.

The Passage without Barrier

The unanimously interacting space-time charged with the harmonious flow of the boundless dimensionalities effectuates an engineered landscape in which frictionless mobility of the constituting elements remains self-contained. Likewise, another popularly savored conceptual object that can also demark the territorial flexions is the geometry of the Klein bottle—elucidated by German mathematician Felix Klein (1849–1925), well known for his mathematical acumen, especially in advanced fields of algebraic geometry. This artistically gratifying structure can be constructed from a penetrable elastic sheet that is first rolled to merge the boundaries along its length, followed by passing one end of the resulting cylinder through the side of the cylinder, where it travels all the way to merge with the other end (figure 6.1). Whereas the Möbius strip has one surface, the topological form of the Klein bottle, which represents a two-dimensional manifold, is altogether devoid of a surface occupancy—the configuration is nonorientable. The nonorientability of the resulting object is an example of how the continuum of distinct surfaces renders the effectuation of an autonomous space that nests warped along its own shape. The reason for its nonorientability stems from two causes. First, obviously, we have warped both the original margins—the length as well as the width. Second, the act of traversing—the passage of one end of the cylinder across its own build—would be possible only if the texture were uniformly transformable. Hence, here again, the demarcation of exterior and interior melts away.

Along these lines, one more profile that draws attention is the time-honored faddish form of torus, which also depicts the continuity of the constituting elements. In this topological fabrication of numerical coordinates, the hypothetical mold again is structured of an elastic sheet. The fringeless spread is obtained first by merging the edges along the length, followed by cementing the resulting cylinder over the rims of the original width (figure 6.1).

Whether we are assessing the continuum by the complex manifold of Calabi-Yau or the simple two-dimensional folds of a Klein bottle and torus, these ravishing mathematical maneuvers can throw light on how a unanimous conformity can modulate physical reality for the flow of dimensions as well as the entirety of space-time; this is in recognition of the sheet like structure (the byzantine mathematical texturing of "D-branes"[7]) of space-time that is appearing in the contemporary formulations. I will come back to the backbone of this membrane-like absoluteness in reference to the beating of the parallel universes. The kindred features of barrierlessness and self-containment that come about via the prescription of the fundamental principles come to manifest beyond the materiality of sight, in accordance with the proclamations of the quantum data and impersonations of algebraic tones.

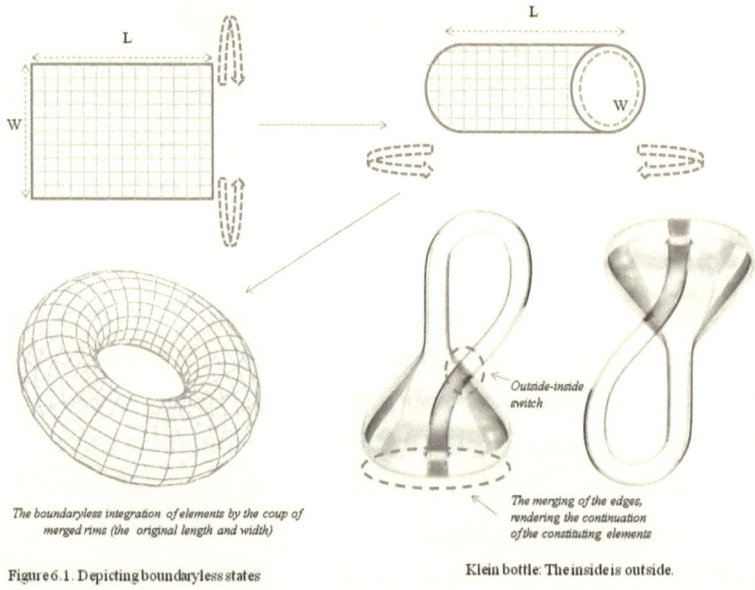

The boundaryless integration of elements by the coup of merged rims (the original length and width)

Outside-inside switch

The merging of the edges, rendering the continuation of the constituting elements

Figure 6.1. Depicting boundaryless states

Klein bottle: The inside is outside.

With these depictions in hand, moving into the boundaryless-ness of space-time may begin to expose the streaming of the absolute space, but we must blend the reality of scientific views with the accuracy of our judgments into a singularity to possibly see that.

Rationalizing Boundaries In the Encyclopedic Perspective of the String

The unobstructed river of succession that the traceries of Klein bottle, torus, and the Calabi-Yau manifold impart narrates a landscape that is not just self-contained[8] but also uniformly dynamic under a single overarching structural scheme of synergistic regulation.

The steering of the colossal amplitudes of the cosmic landscape by manifold dimensions that we learn in current physics in retrospect poses a grave mystical eagerness as to where we ourselves stand in the limits of this convoluted plane. As noted earlier, the most commonly held explanation for the existence of the multifold dimensionality is that the six dimensions, out of the total of ten, stretch only infinitesimally and remain curled up in a form that is analogous to the Calabi-Yau manifold.[9] This judgment sprouted from the earlier idea of curled-up dimensions asserted by the model of Kaluza–Klein theory, in an attempt to unify gravitation with electromagnetism (in the year 1921), by German mathematician and physicist Theodore Kaluza (1885–1954).

The thing is, in connection to the way we perceive the universe, if the six dimensions were to be fixed as minutely compacted, independent of the three extended ones, then any object we see not only exists in the extended plane but also spans the compacted venue, precisely because, in accordance with this explanation, every point of space-time that sweeps into the extended dimensions spans those six miniaturized dimensions as well. Hence, for instance, every electromagnetic wave would span all the teensy dimensions as well as the extended ones—every speck of space-time, from cosmic to atomic bodies to our own occupancy, would sweep through all nine dimensions, the extended as well as the curled up, with equal certainty. This also implies that we ourselves, as well as all the objects we bump into, instead of thriving in three dimensions, reap into the tangle of nine dimensions—what a skewed portrayal (not to mention that the idea of the universe swinging in such a weaved tangle instinctively appears to be pretty far-flung). In contrast, our senses as a matter of fact attest to the steadiness of three dimensions—or perhaps an error in understanding? This eerie scenario brings on a further creepy incongruity—does our experience of life occur via the extended plane or by the means of the curled-up one?

In the design of space-time, the setup of dimensions is again just an outcome of how mass-energy dynamics under the influence of universal forces remain unanimously optimized. This is to say that the inauguration of dimension is an outcome of the optimized forces and not an arbitrary result independent of those very rules that underlie the fabric of the matter; hence, the view of distorted space-time to reconcile the theoretically emerged number of dimensions might be an extremely unrealistic speculation.

The emergence of such an anomalous picture takes form as soon as the prongs of space-time are classified and segregated as autonomous occupancies. Our perception of an object corresponds to the establishment of electromagnetic radiation between the object and the eye, and the advent of this connection directs the articulation of the spatial dimensions—which we numerically assign as three, though in the full-dress image of the universe they exist transcendentally, boundlessly amalgamated into an indivisible connectivity (figure 6.2). Therefore, the theoretical emergence of a six-dimensional transcendental Calabi-Yau manifold in the landscape of reality may then just emblematize the accounting of dimensions as discrete extensions in the parameterization.

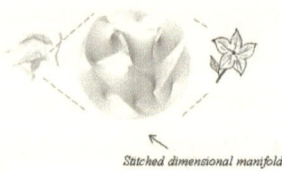

Stitched dimensional manifold

Figure 6.2. The trans-dimensional blur in the establishment of sight and space-time. The presence of the mathematical manifold here highlights the display of the sewn dimensions, where the strands of dimension do not radiate as discrete extensions but rather tie up to exist in continuity.

Manifold image (Calabi-Yau manifold) credit: A. Hanson

Although demonstrating an excellent fitting of physical phenomena into a harmonious whole, the legitimacy of string theory stands on unverifiable grounds, mainly for us not to be able to detect them directly. The imperceptibility of the string, in fact, is the main reason why the size of the string is presumed to be so miniscule, whereas, in accordance with the uniformity of universal forces, the cosmic arena must dress into the strings as much as the quantum region does. Reconsidering the hazy architecture of the string in hindsight, however, does shine an intuitive resolve toward envisaging the string manifestation to be real. Rather than considering the strings as being discrete entities, if we can regard them as a precise design of interacting resonances representing space-time, where we perceive life from within such traceries, we may be able to view the compeller footing of the string phenomenon in capturing the landscape of reality. Besides, the idea of string being nothing more than a phenomenon solely of resonation would further grant us the way to cement the string narrative into the cosmic paradigm, in catching the grand design.

Now let's try to capture the boundaryless state as evinced by the observer.

The Seal of Boundarylessness in the Window of the Observer

Engineering the topological space of the torus and the Klein bottle entails tying up the two dimensions into a single circuitry and making the circuit system interminable (figure 6.1). The state of implicit dimensionlessness, or the sway of dimensions in transcendence, professes that not just the two geometrical extents but also the third extension of this three-dimensional décor remains in continuum or warped. Envisaging the warping of the third dimension might appear bafflingly tricky. We saw a few topological forms that emerge by concealing the length and the width; however, the concealment of the depth as well (figure 6.3) would render the disappearance of the three-dimensional plane of reference itself—three-dimensional mold tucked away in whirls of the higher-dimensional field. In accordance with physical principles, this burying of the dimensions in toto would correspond to the warped causality fashioned by the potentialities of electromagnetism and gravity, by which we ourselves exist in relation to the rest of the physical terrain.

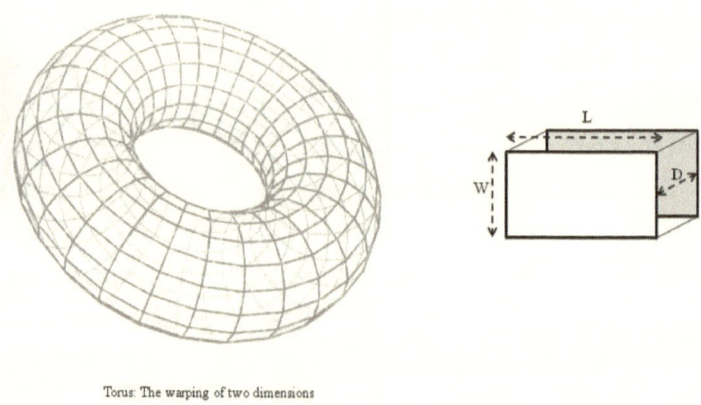

Torus: The warping of two dimensions

Figure 6.3. In case of three-dimensional warping, the depth, along with length and width, will assimilate into the same homogenous totality of boundarylessness.

The warping of all the dimensions, the gesture that equates to the implant of the omnipresent relativity, in and of itself would be a natural outcome if the physical forces were to flow harmoniously. And in such a schematic, the enshrouded reaches of the ultimate space-time, the one that embodies our own existence, becomes a self-contained occurrence. Any physical entity, including the perceptive presence, cannot exist independent of the furthest space-warp, mainly because an embodiment cannot oscillate independent of the harmonious channeling in the flow of the physical forces. The tucking away of the dimensions must associate with a field that would circumnavigate the three-dimensional plane, if the stretches of dimensions were to remain in continuum, and thus lie embedded in relativity. Lying independently of the elemental continuum, from the perspective of the higher vista, the cocooned materiality would then stay established as a transcendence of the three dimensions.

The identified occupancy of the observer from the premise of the relativistic outlook to the delineation of the quantum world, though, has been scientifically acknowledged to be a peculiarity in the manifestation of that which is perceived: the much-desiring enquiry of the positioning of this observer in the scientific picture of the whole has perhaps not been made with diligent sincerity. For instance, the positioning of the observer in relation to the accelerating universe, although it bears decisive significance in perceiving the overall scheme, has not, by and large, been esteemed genuinely pertinent in the current tracks of scientific ratiocinations.

Whether we are looking to consolidate the deluge of scientific verdicts or to take in what realized beings point to, the bespoken positioning of the observer on the inverse side of space-time fittingly construes a picture of a synchronous system that embodies our own inhabitance in the weave of the mass–energy dynamics—the subject to which I shall return in conjunction with the correlating topic of the role of the observer in the coming about of the parallel universes. Here it may help just to add that emerged heterogeneity of the observer in scientific outlooks occurs not because there exists a multitudinous array of observers but because of the established mosaic of mass-energy forms, where each "mass–energy outline" identifies the presence of the nondual observer beyond its own physical extent. Thus the observer, although remaining nondual, at the plane of the matter is deceptively shown multifariously. I have attempted more clarification on this with deeper grounding in later chapters.

The stretch of the observer that spans about the ensconced whirl of dimensions is the expanse in which all that we physiologically notice, including the clamp of the self, converges at the juncture of self-realization or enlightenment.

The borderlessness, in turn, is an essence to a foremost behavior of space-time: the across-the-board flow of transformations.

Prerequisite of Transformation in the Working of the Dynamic Universe

The algorithmic modulation of Möbius transformation gives us a way to conceive how the manifested fluidity in the textural core of the universe that the wisdom of physics conveys comes by. Such a mathematical portrayal insinuates the underlying transformability in the sweep of space-time, dictated by a single set of principles.

Immersed in the conception of wave–particle duality, the feature of transformability is pivotal to the flow of time and the latch of space. The wave and the particle are two different ways of looking at the same embodiment, and so the space successively transforms in echoing the flow of time. We will explore the crux of the transformability in the machination of the reality with greater detail in chapter 10. From the viewpoint of the observer, the game of mass–energy schematically orchestrates as a unitary system in which before the eyes manifests the worldly plane, and behind them, the inner or the mental plane, for any given point in time (figure 6.4).

The emblematic Möbius transformation of the homogenous sys-
tem, along the same lines, correlates to the plaster of π and ϕ that
is equivalently uttered in the casting of the subliminal and the
materialistic spans—the two disconnected systems that deflect a
supersymmetric collage from the standpoint of the sensitivity of
the observer (figure 6.4).

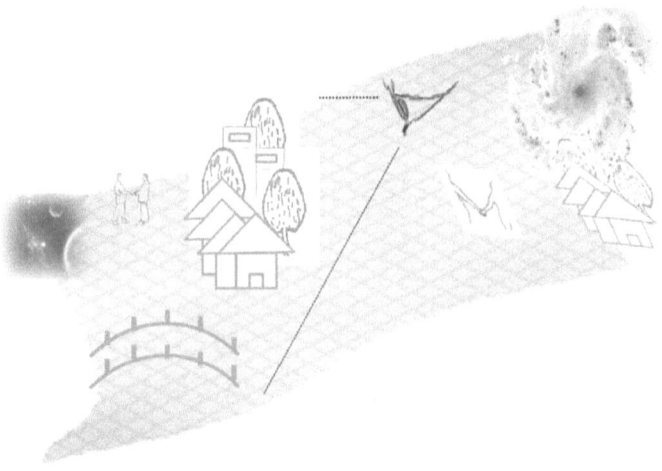

Figure 6.4. From the perspective of the pure observer, manifested alongside the furls of space-time, the specks of
the worldly and the subliminal planes are parts of the same spread, where only the prior is captured neurally.

The nondual watcher not only senses the subliminal span and
the ocular functionality that plants on that span but also optically
witnesses the worldly frame, stretched analogously along the inner
reach.

We should have sufficient grounding to now attempt to unify
the purely observing attribute, which we remain unfamiliar to, and
our perceptivity, which we have a good sense of, toward bringing
the scientific declarations and our own placing, and evolution, under a single overarching flow.

CHAPTER 7

||

Are We Dreaming or Awake?

*Cues from the Set Theory: Constructing
the Universal List of All Lists Using
the Voice of Empty Space*

The Hint in the Mathematical Paradox

British mathematician and philosopher, Bertrand Russell (1872–1970) contrived an enigma that springs up in the logistical understanding of a mathematical order, the decipherment of which seemingly rests in the unraveling of how the physical is sculpted with the metaphysical in inducing the all-inclusive fabric.

Under the rubric of the set theory, this quandary, popularly known as Russell's paradox, sets in over our insistent need to determine whether a set itself is a member of its own set.[1] As it goes, in this tale a barber in a town shaves those who do not shave themselves, and thus, in this town, there occur two types of sets, one comprising those who go to the barber and the other those who shave themselves. The contradiction emerges when we stand in need of knowing which set the barber belongs to. The barber shaves himself; would he be a member of the self-shavers, or does he belong with those who go to the barber? The resolution is straightforward if the barber resides out of town or if barber is a woman; then the barber does not belong to either set. However, the barber residing in the same town and being a man are the axioms that are included in framing the riddle, setting up a puzzle that cannot be clearly addressed, until we find a way to conjoin the two sets and yet somehow keep them distinctive.

The clear texture and the nitty-gritty of the preceding problem may become relatively hard to grasp, especially for the noise generated in seeing the two sets distinctively in reference to the barber. Along the same lines, however, yet another perplexity can provide a clearer appeal to the grappling situation—a set standing in need of being a member of its own set—in the custom of the set theory. Again voiced by Bertrand Russell, a twist of "a set of all sets" asks for the drafting of an all-over set that contains all the probable sets—composing a juncture that, in retrospect, obligates that this all-over set be a member of its own set. For instance, the making of a universal list that contains each and every list there is requires that *it* should subsume *itself* too, or else it cannot qualify for the status of being the universal list. And thus, in the assemblage of universal infinities, a set that is declared to contain all of the number sets should contain itself too. To appreciate the substantiality of identifying a set as being a member of its own set, and realizing its instrumentality in the real universe, we need first to take a breezy stroll through certain core facts of the set theory that refer to the strategies involved in the packaging of infinite sets.

Building Numerical Sets to Telescope Infinity

The description of infinity has been a delectable subject of mathematical musings ever since ancient times, but it was largely the renown of the mathematician, physicist, and astronomer Galileo Galilei, who inspired the ways to chart the nature of actual infinity. And it was the philosopher and mathematician Bernard Bolzano (1781–1848) who strategically brought into the open the idea of describing infinity by the medium of the arithmetical set, laying out the first thoughts on the synthesis of the set theory. The rigorous structuring of the set theory, however, came in the work of Georg Cantor, and through his collaborative efforts with German mathematician Richard Dedekind (1831–1916).

The discipline of the set theory addresses the methodized categorization of mathematical entities and the axioms and logics that come to play in the setting up of these assemblages. The same bent of arguments was utilized by Georg Cantor in the batching of the number limitlessness toward understanding the actual nature of infinity. The tactful systematization culminated in the appearance of not just one but several (in fact infinite) stages of infinity, arranged in succession. The grades of infinity were denoted by the sequential arrangement of the Hebrew word aleph, $\aleph$, appearing as $\aleph_0$, $\aleph_1$, $\aleph_2$, $\aleph_3$, and so on.

The underlying principle employed in formulating the distinctive sets of infinity was based on the distinguishing counting profiles that the different types of numbers exhibit. The independence between infinite sets was determined when the bijection—a mathematical term for assessing one-to-one correlation, or exact pairing—could not be established between them. Thus, a particular infinite set would contain only those number types that can be matched bijectively. For instance, utilizing this postulate, Georg Cantor showed that the natural and rational numbers (addressed as countable) fall into the same set of infinity, while the real numbers (the uncountable ones) belong to a higher order of infinity.[2] The types of numbers that sequentially pair up were argued by Georg Cantor to possess the same cardinality—a standard designation in the set theory to indicate the quantitative value, in terms of the number of elements, for a given set. Thus, under the edict of the set theory, the sets with different cardinalities would convey that there are different numbers of elements present between them—or different numbers of elements are decisively projected in the case of infinite sets.

For instance, set I {A, B, C} and set II {B, C, D, E} have different cardinalities because set I has three elements, whereas set II has four; Set III {F, G, H} and Set IV {I, J, K, K} possess the same cardinality; K repeats itself in set IV, and within the convention of the set theory, it is only the peculiarity of the element that is accounted in establishing the cardinality of a given set; a repetition of a number type is disregarded. Therefore, no matter how many times a particular item repeats itself, it marks the presence of just one element. Just as the previous two sets I and II declare disparate cardinalities because they do not connect bijectively, the two infinite cardinals ($\aleph_I$ and $\aleph_{II}$), in the same way, stand for a dissimilar number of elements, for they do not display an orderly alignment between them.

The infinite sets themselves are inexhaustible, or there is no greatest infinite cardinal set; any level of infinite cardinality is succeeded by the display of a higher order of cardinality because of the potentiality of the power-set—a truly suggestive exposition that we will come across shortly. The occurrence of ceaseless self-sprouting in the manifestation of infinite sets brings forth a quandary of how to devise a universal set—or, with respect to the nature of reality, how to visualize a universal conglomerate—an essentiality toward understanding the uniformity in the alliance of physical laws. The decryption may lie in pinning down a mode by which a set turns to become a member of itself. Whether we look to crack the barber's paradox or to envisage an all-embracing synchronicity in the description of reality, the configuration of the universal set must be such that it arbitrates, in some coherent way, to make the universal set emerge as a member of its own set. The revelation of such a scheme is uttered in the schematic façade of space-time, when viewed as a marvel that occurs only in conjunction with the span of the scopic observer—we will see how in a little while.

In connection to the concocted paradoxes of the set theory, the subsequent axioms set forth by mathematicians such as David Hilbert (1862–1943) and Ernest Zermelo (1871–1953) patched up many of the loopholes in which Bertrand Russell's work on logic and foundational thoughts on infinite sets played a critical role, and on which betterments were made by the well-recognized American logician Kurt Gödel and skillful mathematicians such as Abraham Fraenkel and Paul Cohen. However, in the real world, a universal set of all sets must be a member of itself because if it is not, then this would imply (1) that one set is left out from the subsumption and therefore the set cannot be called the universal set, and foremost, (2) that the sustainment of something extramural is insinuated; even if that something happens to be merely an availability of an empty space, its very existence beyond the bounds of the universal set contradicts the legitimacy of the universal set. Thus, beyond the margins of mathematical bewilderment, the stride of defining a universal set appears to be a bona fide step in contriving the factual picture.

Here the machination of the power set may help us to gain an understanding of a universality that remains all-inclusive.

The Power Set

The conception of the power set, again furnished by Georg Cantor, alludes to a bundle that comes to appear when all the elements of a given set are placed in all possible combinations, including their own presence individually. The powering of the number of elements, thus, confers the power set a higher cardinality, in comparison to the commencing set.

The set {a, b, c} thereupon fudges together a power set of the frame {{a}, {b}, {c}, {a, b}, {b, c}, {a, c}, {a, b, c}, {∅}}, where {∅} denotes an empty set, which, according to the power set prescript, must string along with the rest of the members—a decisive norm that I will come back to shortly. The elements of a power set are known as subsets. Noticeably discernible in this mixture, the quantity of elements lodged in the power set is exactly the base 2 exponentiated by the number of elements that are present in the original set. In this example, therefore, the set {a, b, c}, which contains three elements, furnishes 2^3 (eight elements, listed earlier) in its power set. Likewise, Georg Cantor showed that the continuing chain of infinite sets remains connected by the following equation:

$$\aleph_1 = 2^{\aleph_0}$$

Here the power set $\aleph_1$ presents higher cardinality (the number of elements) by the magnitude of base 2 exponentiated by the number of elements in $\aleph_0$, the set of the preceding cardinality.

Hidden in the schematic façade of the power set is the working of binarity—a graphical symmetry between being present and being absent for every given element that comes from the precursor set. Each of the elements possesses equal likelihood of emerging and submerging in the cluster of the subsets. The set {a, b, c} develops into its power set by a combinatorial assemblage where the singletons {a}, {b}, and {c}, the two-digit combos {a, b}, {b, c} and {a, c}, the triplet {a, b, c} and the empty subset {∅} collectively feature two modes in equity (for any given initial element): present and absent (figure 7.1). In the power set of the eight preceding subsets, the element "a" makes its appearance four times as well as going hidden four times; so do the other two elements, "b" and "c." Similarly, a power set of the set {a, b, c, d} will constitute 2^4 (16 elements): every element is held displayed as well as kept hidden eight times; the coming-into-sight and going-out-of-sight equilibrium is an outcome of exhausting all element combinations in the coming about of the power set. The elements collectively induce a twofold symmetry, which is precisely why, in formulation of the power set, the exponentiation base is always 2 (see the previous equation).

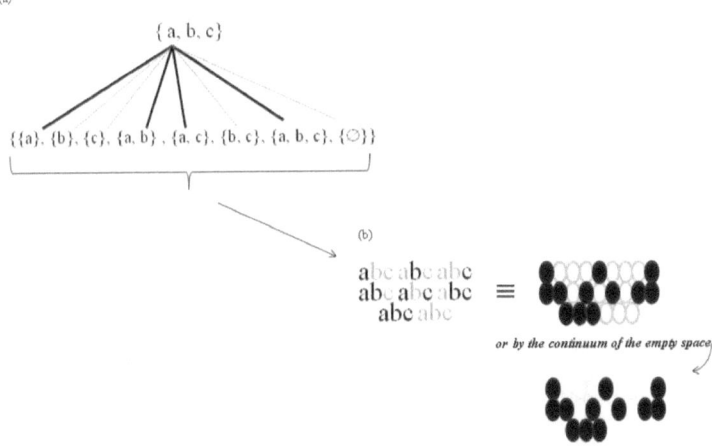

(b)

abe abe abe
abe abe abe   ≡
abcabe

or by the continuum of the empty space

Figure 7.1 A power set. (a) Element *a* is present four times and absent four times (bold line, present; dotted line, absent). Elements *b* and *c* portray the same binaries. (b) Power set imprints the counterbalance between occupancy and absenteeism. Faint letters denote absence. The vacant spots window the continuum of the empty space from behind.

Now let's see how we can perceive a larger picture by the ploy of power set.

A Set Mediating to Become a Member of Itself: The Radical Gesture of {∅}

Underlying the machination of the power set is an attribute that is the most pivotal not just for the strategic fall out of the power set but also for seeing the delicate connectivity between the language of the set theory and the nature of reality—the hunch that led to the paradoxes at the outset. Procured in twofold symmetry, the power set asserts that its subset, which contains all of the elements ({a, b, c}), must ride along a countering subset with a complete absence of elements, figuratively denoted as {∅}. Not blatantly evident for its incisive function, the empty subset {∅} is core to the mathematical texture of the power set and is cardinal toward decrypting the reality by the norms of the set theory. Every so often, obliviously addressed as vacuous truth in the texts, the portrayal of the empty set is literally *the* most suggestive hallmark of this mathematical framework.

In the power set {{a}, {b}, {c}, {a, b}, {a, c}, {b, c}, {a, b, c}, {∅}}, every subset, denoting an element, is taken to be of identical value in the assignment of the cardinality, and therefore, the subsets {a} and {a, b} remain equivalent for their impacts in the accounting of the cardinality. The singles and combos stay on equal footing, for all they insinuate is the countable individualities (figure 7.1).

The bifold equipoise of "show" and "hide" in the design of the power set is revelatory of a suggestive detail, accounting that into the synthesis of the power set a design can be viewed in which a set *can* be a member of itself. Within the habitat of binarity, where an empty set {∅} inhabits the set, a set can be reasoned to be a member of itself. Let us see how.

In the crisp composition of the power set, the presence of a regular element signifies the presence of a property. In the real world, this property can be anything: a physical object, from a quantum particle to a cosmic galaxy to the entire universe; a man-made object, a creativity, an artistry, an act or a behavior; an emotion, a feeling, an imagination, a mental image, a thought, or a perception; a concept, a logic, an understanding, a theory or proposal. Anything that exists within the bounds of physical and mental makeup would adumbrate the show of an element. The empty space {∅}, in contrast, bespeaks a propertyless denomination, and in agreement to the blueprint of reality that mathematics emanates, the inclusion of this attribute-lessness flashes a state of oneness, subsisting beyond properties. The establishment of this unhindered continuum is needful in figuring out the inconsistencies that develop in the framing of the universal set and in defining the mechanism the power set comes by.

This is to say that in the spotless draft of the power set, the empty space bestows a continuum through which the elements are made known for their presence and absence (figure 7.1). The empty space stays screened at the regions where the elements are declared. While the crystallization of the subset {∅} is a consequence of the presence of {a, b, c} (the subset that has all the elements) in the establishment of the bifold symmetry, the empty space ∅ by and large acts as a continuum in the plot of the power set. Whereas the bracketing of the empty space {∅} symbolizes counterbalancing the subset of {a, b, c}, the overall appearance of power set symmetry itself is pillared on the continuum that is fixed by the ploy of the empty space—more details on this in later sections.

A set, thus, in an enfolded continuum of its power set, *can* be seen to be residing as a member of itself. Although the presence of subset {∅} signifies the absence of subset {a, b, c} in the bifold symmetry, the realization that the wall-to-wall presence of the empty space is inherent in the systematization of the power set proffers a concatenating scheme; in the layout where the empty space renders continuity, the set {a, b, c} is also the subset {a, b, c} (figure 7.1). The set becomes a member of itself.

The Space Pretzel in the Back-Cloth of {∅}

There *can* be a universal list that has its own copy because in its power set its own presence is encased by the executed empty space, overdrawn beyond the elemental motif; thus the subsets, epithet elements, remain warped, whereby the subset that has all the elements from the beginning set, and thus imaging the germinal set—{a, b, c} in the earlier example—becomes a part of the all-inclusive weave (figure 7.1). It is because of the functionality of the empty space that the metrical structure seen from within the power set could be identical to the structure perceived from outside of the power set, for, under the expanse of the empty space, there remains no interior–exterior differentiation. The empty space renders the interior–exterior continuum feasible, making the "universal set" akin to the "universal subset" that dwells within it.

Working out the puzzlement that involves a human subject is a relatively trickier matter, for it involves the personification of the mind factor. The barber feels that he is the barber, and so do the rest in town; they feel that being-the-barber is a permanent state rather than a passing functionality. The bafflement of requiring a being to be and not to be (the barber) melts away as soon as we consider that being-the-barber is only a functionality, and by the same token being-the-barber and not-being-the-barber are just two different dynamic elements, where the personality that is tagged as the barber, or not the barber, in the purest reality is continuing as the back cloth of the empty space, beyond the interactive dynamics—paralleling the discrete elements of the set—that go on in the town.

In the plot where the hindmost reality of the vitality that is tagged as "barber" is the sweep of empty space, the operability "barber" comes to be the set of itself; thus, in the preceding paradox, the functionality "barber (shaving others)" is placed in the empty space the same way the functionality "nonbarber (shaving himself)" is. Being-the-barber and not-being-the-barber solely stand for two different characteristics, akin to the nested elements (or the subsets) in the equilibrium of the power set. The symbolical enclosed empty space—the subset $\{\varnothing\}$—is identified as an element purely because of the confinement imposed by the brackets. The emergence of the subsets, encrusted with a uniform symmetry, rests on the foundation of the empty space continued beyond brackets; we will notice how in the next chapter. The empty space can identify itself as any particular element; the empty space, by the same token, can be identified as an entire "universal list," or as a subset ({the universal list}) in the power set.

The perfect tracery of the power set, brought by the equitable roles that elements and the empty space play in this mathematical apperception, shines as a perfect specimen toward contemplating the texture of the real-world universal set, and how our own positioning might be enmeshed in this overall pack. The real-world set would be an exhaustive assemblage of the elements from the dominions of the universe and the mind, surging pattern-fully in coordination with the background of the empty space.

The Reality of the Dream or the Dream of the Reality

Relating the previous mathematical assemblage to reality brings on a need to contrive a justifiable role that the empty space enacts in the manifestation of the totality, based on the clue that the cognomen "empty space" is not just symbolic but also bears a clear stance in the fruition of reality. The canvases of the quasi-reality that we descry when we are asleep relate in various ways to the universe with which we rendezvous while awake. In reference to our cognitive perceptions, the reign of the universe and that of the mind unitarily present themselves as a single congregation. The integrity of the two reigns is brought on by the unshaken visionary sense that continues to be in effect whether we are asleep or awake. Besides perceiving the universe by eye, our visionary sense occurs at yet another distinctive level—at the level of dreams while we are asleep. This visionary sense, the watchability of all of the nitty-gritty, remains independent of the domains of the worldly and subliminal planes, as it is this visionary sense that detects the universe, the conceptions, the thoughts and emotions while awake and the dreams while asleep. Thus, "we" as "beings" in this ultimate sketch recognizably continue purely as onlookers, fully perceptive of the complex dynamic arena. We know that we are the watchers of the dream as well as that of the mind's activeness, and so, in absolute purity, we *are* that visionary sense, independent of the elemental characteristics that we have pictured in the mathematical set. By the same token, the mathematical empty space that is essential in whipping up the power set corresponds to the pure visionary sense (the watcher), which is a core ingredient in whipping up the utmost fabric. The fabric that includes our nature and all the paraphernalia that tag along with it.

By the context of the set theory, the empty space (the watcher) can identify itself as a worldly or psychical element. But beyond the norms of identities remains a pure watcher that comes upon dreams, universal hierarchies, thoughts, and intellections. When identified, it signals a man, a barber, a personality, any worldly character, or a mind state for that given moment (figure 7.2). Beyond identification, it continues as an empty space located beyond the grid of properties.

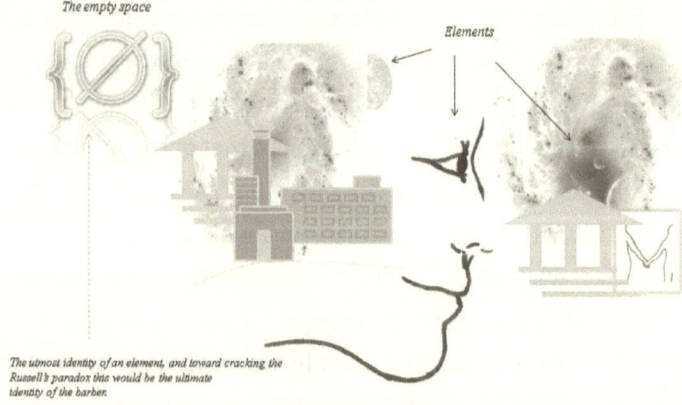

The empty space

Elements

The utmost identity of an element, and toward cracking the Russell's paradox this would be the ultimate identity of the barber.

Figure 7.2. The span of empty space goes on alongside the dynamics of the elements, constituted by worldly and subliminal parts.

Phenomenally appealing to note is that from the viewpoint of the utmost identity—that is, the watcher—the world that is seen and lived in is no better or worse than the dreams that are encountered while asleep. Elements of the dreams and the world breeze in to subsequently pass away, restating that the display before the eyes and the one behind them stand for similar relevance, or irrelevance. Connectively captivating is that in reference to the worldly encounters, the dream state while asleep at times appears to be more vigorous, for while dreaming *we* melt into the dream so deeply that we remain relatively oblivious to our own worldly individualities (figure 7.3); the certainty of the grandiose reality was de facto a delusion of the hazy dream become known only upon awaking.

Figure 7.3. Dream with the eyes closed is as real as the world with the eyes open.

In bounds of pacific sleep, our own identities efflux as a part of the dream stage—an identity that in many ways is different from the identity present in our worldly associations; the worldly character is asleep. The phantasmal sleepy character is then unknowingly the bystander of the dream—a dream that has the self woven into the play. Thus the staged phases of awakeness and asleepness in-volve the shifts in the nature of identity from the worldly performer (when awake) to the watcher of the dream.

In recollection, from the viewpoint of the *watcher*, the awake-ness would signify collaborative dynamics between the subcon-scious and the tangible components, and consequently, with the surge of earthly activity, the watcher goes fully enveloped and we stay afar from acknowledging our own ultimate complexion. And, in the scenario where the watcher resides fully concealed, we would stumble to regard the dreamy land of physicality as a spread of thickest reality.

Thus, reflectively, beyond the tags of identities, both the physi-cal arena and the imaginary canvas congregate into an unabridged power set, radiating by the rightful regularity.

CHAPTER 8

δ

The Arrival of Higher Dimension

The Universe Needing a Higher-Dimensional Field of Continuum

Still the Inspiration of the Power Set

The deft medley of the elements in the rigor of the power set imparts yet another subtlety—an affirmation on the accuracy of the correlation between this mathematical conceptualization and the demeanor of the actual existence. The binate order in the tapestry of the power set carries a harmonic layout in which one half assumes the positive and the other the negative of the same image (figure 8.1). In the power set of the set {a, b, c, d}, which would contain 16 (2^4) elements, every primordial element lights, as well as darkens, eight times, producing a layout such that half of the power set becomes the graphical inverse of the other bisection. Toward understanding our own positioning in the universe, the symmetry of the power set furnishes a simple graphical imitation—indeed, after which it is the code of physics and our interpretations that would endorse and strengthen such a picture, and I will be discussing that in due course. It may help, though, first to consider the design of real inner–outer conformity by the gesture that mathematics offers.

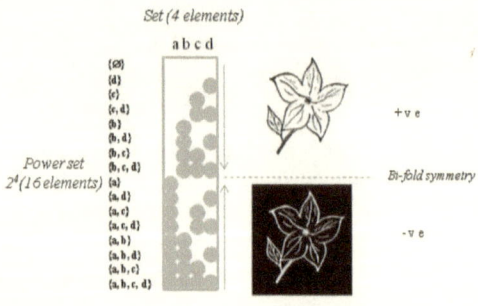

Figure 8.1. Power set symmetry. Every original element is present eight times and absent eight times. Half of the power set is inverse of the other half. Positive and negative are created in concurrence.

The symmetrical landscape of positive and negative can symbolically relate to the erection of a stubborn mirroring that is secured between the substantiality of the tactile universe and that of the mental deportment—a juxtaposition that is essential if we are to conceive a unitary plan that imbeds our own presence equably. Let us muse over this reflective equilibrium that in all its reasonableness advocates that the alliance between the observable and the inner arenas is all the more constitutional in the fallout of reality.

The pairing of the opposites in the in-unison plot of the power set is combinatorially dictated by the turning up of the combinations of the original elements and the emergence of the empty space behind them, when these inceptive elements go missing. For instance, the subset {d} (see figure 8.1) doesn't just signal the dwelling of the basic element "d"; it rather reports that the spatial occupancy of the element "d" is seated after the instances of the other three initial elements, "a," "b," and "c," go missing, which in consequence causes the emergence of the empty space, equivalent to the occupancy of the elements that are currently left behind. The subset {a, c, d}, therefore, de facto lies as {a, empty space equivalent to b, c, d}; the subset {a, d}, the same way, equates {a, empty space corresponding to b and c, d}; the subset that masks the stretch of the empty space altogether counteractively accompanies a subset that is all void, {∅}. An easy-to-understand pictorial elucidation of the counteractive coupling can be seen on the website Math is Fun.

It is also obvious that parallel to the overall bifold reversibility, the feature of polarity occurs at the level of subsets too; every subset is accompanied by an inverse image (figure 8.1). If a subset comprises three basic elements followed by an empty space, the concomitant negative subset would flash the lack of those three elements followed by the appearance of the element that is missing in the counteractive subset.

The coordinative layout of positive and negative instills the potentiality of self-annihilation in the sketch of the power set; folding the power set along the axis of symmetry and aligning the two halves (figure 8.2) makes the structure of the power set itself disappear, for the outline of one half that stems from the rise and fall of the elemental space, in its entirety, offsets the other half, which projects the same design contrariwise. In view to the gross structure of the power set, it is not that one particular half stands to be positive against the form of the other; it is just that the nullifying images are the outcome in the scheme of the power set. The existence of the positive is coupled with a copy of the negative; the contrasting images coexist. Here positive and negative both imply the same paraphernalia, cooked up by the empty space and the elements, projected in two exact opposite ways (figure 8.1).

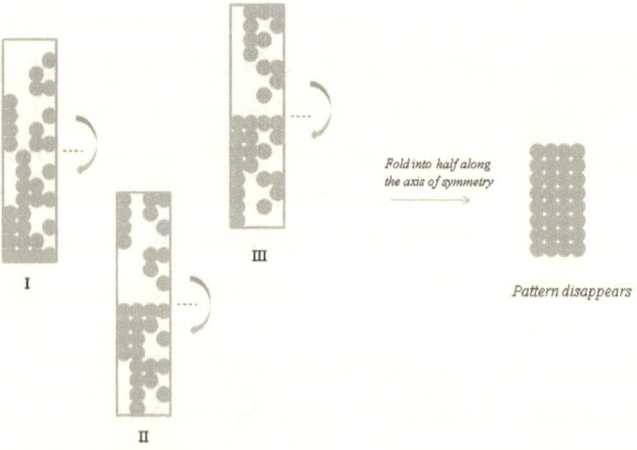

Figure 8.2. The power set. Positive and negative are effectuated in concurrence, evoking the feature of nullification.

Having acknowledged the mirroring gesticulate of the mathematical embodiment that fruits only in the ambience of the empty space, the ditto event of the symmetrical manifestations in the real universe may not appear farfetched when we, with all expository notations, account the dwelling of the empty space (or the watcher, as noted in the preceding chapter) as equally potent in the sketch of reality. The embracement of the watcher consequently allows us to assimilate a unifying rendering that teams with all of the intellectual belongings within a single across-the-board framework.

Whether we recognize the dire necessity of interlacing the matter with the antimatter, the spread of energy in the shadow of mass, the flow of wave under every speck of the particle, and the cryptic incongruity between the quantum and cosmic realms, or recall that the mosaic of the mental elements infiltrates as a sturdy ingredient rather than as an arbitrary triviality in the coming about of the greater reality, the hammer of symmetry is assuredly seen glimmering in every possible way we peek into the grand conformity. And it is the same counteractive bi-fold that emerges in the ambience of the watcher, from the fallout of the two distinctive images of the same scope. One image as the tactile universe and the other as all-encompassing breadth the way we grasp quantum mechanically. The all-encompassing breadth carries the subliminal elements, making the tactile universe in symmetry with the mental plane for every time point, the way the mathematics presents us. Indeed, the mold of such a symmetry would involve the participation of both the emotive and cognitive elements.

Why is such an all-abounding symmetry procured? Why such a counteractive framework deeply permeates the grand scheme? Why the feature of annulling seeps the existence? How do we explain the erected parallel universes that physics augurs in the model that insinuates a single overarching watcher? These are all the most transparent inquests, and I will come back to them in the next few chapters.

The Existence by the Coexistence

The layout of the autonomous space-time, which stays warped, remains encrusted by the characters of contrast, in many ways, and simultaneity—the features cued by the analytical endeavors and the plot that emerges from them—and such a schematic ingeniously illustrates how the integration of our own personal elements is procured in the weave of the ultimate amalgam. The scientific schemes and the associated numerical formulations call attention to an illustration where the objects of the universe can exit only when they are accompanied with a copy that vibrates in concomitance, so much so that the theme of existence evolves into the theme of coexistence. All—in entirety or in bits—abides by dualities, come what may. The feature of duality is vital to the beat of space-time, and I have dedicated many sections of a later chapter (chapter 9) to it.

The revelatory gesture of coexistence—the self-sprouting of the mirror images—in the mode of the power set also declares that the concomitant counterpart resides in an independent plane, disconnected from the other fraction. Both the spatial planes image the very same paraphernalia, sewed in contrasting colors, and so their presence in the same plane would nullify each one of them. In the real physicality of this universe, the sway of two such planes, or the discrete three-dimensional fields, entails the presence of a pivot point at the juncture of the two fields that composes an association between them. After all, the two three-dimensional fields are the building blocks of the very same execution.

Relishing the structural layout of space-time that presents itself as the two definitive planes may appear a little far out, especially in the haze of the wavering mind frames that we only cognize but do not optically sense, and yet without hammering in the bulk that lies abaft the ocular, the understanding of the all-across scheme cannot be gained.

The ocular sense by which we discern the universe electromagnetically is the rigor that has the world located frontally and the subliminal makeup set up at its back. Established as a pivot point in the realm of the compounded existence, the five neural holds install that hinge from which the world is acknowledged on one end and the mental makeup on the other; both would be the parts of the same universal design. And from the vantage point of the watcher, the field detected optically and the one sensed psychically would both be parts of an overarching three-dimensional plane, distinctively lodged in congruency; the equilibrium sustains at the uppermost level, at the level that persists as the empty space, the same way the formulation of the power set pitches. By acknowledging our own unwavering sense of watching, we can translucently make out that the nature of the ultimate, in reference to any functionality, is the existence of the unfettered watcher—the witnesser of the counteractive images that the external and the inner stretches transmit. The counteractiveness of the world and the subliminal may not be taken as literal weight-by-weight but rather as the case in which intellectual and emotive bearings play out.

Let us peep a little into how mathematics discourses the structural mechanics that take place in the packing of the two discrete planes.

Mathematical Lens to Spot the Higher Dimension

Tesseract—the term coined by British mathematician Charles Howard Hinton (1853–1907), known for writing science fiction and for his interests in contriving the graphical interpretations of higher-dimensional concepts—commonly known as a hypercube, is a mathematical form that graphically shows how the heap of three-dimensional spaces are stacked in a higher-dimensional span. (A scene in a recent movie *Interstellar* neatly depicts the higher-dimensional order of tesseract.) Specifically, the tesseract alludes to a bold prototype of a four-dimensional space; that is to say that here the two three-dimensional spaces are bundled in a higher-dimensional plane.

The development of the fourth dimension is an automatic occurrence that takes place subsequent to the installation of two or more three-dimensional strata—whether we are addressing the higher compass that embeds the perceptible, the cognitive, and the electromagnetically sensed world in a single ensemble, or a superfold where the parallel universes come to collate into a multifarious reality. The object tesseract identifies the packing of two three-dimensional cubes that literally can transform into one another plainly because these three-dimensional spaces are connected in the higher-dimensional stage. This illuminating depiction, in this manner, comes to be the four-dimensional cognate of the cube the same way the cube is the three-dimensional affiliate of the square—the only difference is that the transformation from the square to the cube involves an alteration in the number of dimensions. While, in the transfiguration into the fourth dimension, or any number of dimensions beyond that of three, the dimensionality, and so the size, of the object does not undergo a shift. In fact, in reference to the higher-dimensional stratum, the size difference becomes inconsequential. A lucid interactive rendering of the tesseract can be found at the website of the Mathematical Association of America, in an article by Alex Bogomolny, or on YouTube. Looking from within the three-dimensional extent, the consociation of the two three-dimensional turfs in the tesseract diagrammatically appear as if one cube is tucked inside the other. However, in this commune, both cubes are identical in measurement and stay separated by the identically sized edges. The delineation of the fourth dimension by the gesture of the tesseract can be appreciated better in animation (easily found on the YouTube): the switching between the two hexahedrons that nest into each other, where one develops into the other, within the arena that registers only a solitary and sequestered three-dimensional surface. The tesseract constitutes eight cubical cells of identical size: six of them stem from the expansion of the six primal cube faces, and the remaining two point to the first and the resulting cubes. In the corporeal expanse, these two hexahedrons can also be equated to the occurrence of the two terrains that are separated in time (figure 8.3).

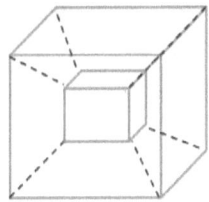

Figure 8.3. Illustration of a hypercube or four dimension. One cube is not bigger or smaller than the other, and one cube is not inside the other. The two cubes can be transformed into one another owing to their connectivity in the four-dimensional space

The hypercube likewise furnishes a scheme of how the parallel universes are structured mechanically, a subject on which we will embark in the succeeding chapter. Akin to the establishment of the concrete and the cognitive realms, structured through the pivot point of the neural perceptions, which crop up in conjunction with the vista of the watcher, the multitudinous universes also converge in alliance with the continuum of the higher-dimension watcher, the way the tesseract outlines. I shall choose to forgo the discussion relating to the culmination of the parallel universes for now; there will be a fresh start on this with a clearer context in the coming chapters.

In the matrix of the hypercube, although each cube speaks for a befitting warped space and comprises its own set of elements, the two cubes still share the eminence of the very same empty space; the premise is foremost in guiding us toward the realization of the mechanism by which the parallel universes spring.

Three-dimensional entities seem to be ducked in the scope of the fourth-dimensional view. However, whether we are speaking of concepts in mathematics and physics or the existence of the three-dimensional world, the fourth dimension is not an extent that pounds above; it simply stays beyond the third dimension, separated from the limits of measurability. The fourth dimension, or any dimension higher than the third, for that matter, states the attribute of coexistence—by which two or more three-dimensional spreads are sustained in continuum.

The Higher-Dimensional Totality Enclosing the Dabs of Conscious Presence

The stellar scheme that embeds our own peculiarities must be extrapolated into grasping the vitality of the parallel universes, the settlement of which is unfailingly announced in quantum mechanical findings. The delineative must incorporate not only the inputs from scientific deductions but also the components of the subliminal, aptly knitted to accord the grand play of the universal parallelism—along the lines of our acknowledgment about the righteousness that comes to play when we embed the components of the mental plane in the scheme that physics portends and mathematics portrays, at the level of the singular universe. In the light of the continuum that shimmers as the watcher in the design of space-time, the associative role of the watcher yet again becomes a decisive stroke in the dynamic fallout of the parallel universes as well—a scheme by which we can envision the presence of subconscious elements not just in the arena of the solitary universe but also in the framework of the parallel ones.

With the utmost status being the watcher (or the empty space in the mathematical lexicon) and not the rippling characteristics of space-time, an association that subsumes the traces of the subconscious and the neurally perceived world would emerge as a package. And by the continuum of the watcher, the transient elements would bob up as the panorama of a unitary universe, for every given time point. The warped power set that insinuates all of the perceptible universe and human factors de facto hymns the cast of an unregimented universe and utters the fixation of a single three-dimensional breadth in the field of the higher dimension (figure 8.4).

Whether we are absorbing the allurement of mathematical paradoxes, figuring out the unification that comes by the ubiquitous implementation of physical laws, or ratiocinating the roles of the subliminal in the framework of reality, the grandeur of the empty space, alias "watcher," must be realized to piece all into a just whole.

The imagery of the internal makeup and the encountered

universe in summation come to be a combinatorial regularity, manifested in near symmetry by the pivots of senses and the continuum
of the watcher. In reference to the watcher, the subliminal factors
and the sensed universe ripen into a solitary across-the-board game
of dynamism (figure 8.4).

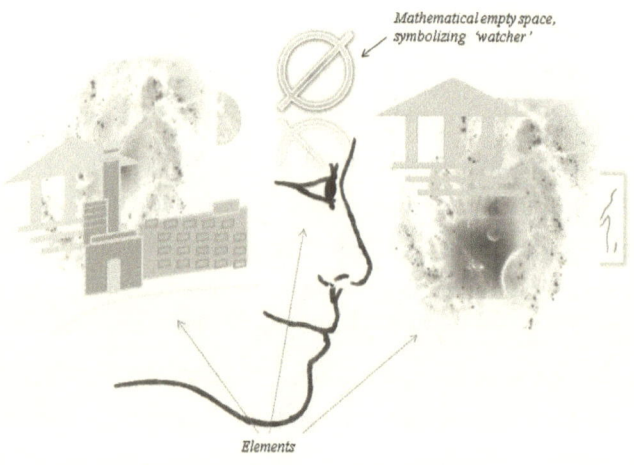

Elements

Figure 8.4. With the utmost identity as the pure watcher, or the mathematical empty space, and not the characteristical elements, every time point signifies a symmetrical dwelling of the palpable and subliminal casts. A symmetry in which cognitive elements play out.

CHAPTER 9

$\forall x P(x)$

The Culmination of Multitudinous Universes by the Sovereignty of Duality

The Empty Set $\{\varnothing\}$ of Parallel Universes: The Nature of Quantum History

The Surfacing of a Concept: The Crystallization of an Intuition

> Thus metaphysics and mathematics are, among all
> the sciences that belong to reason, those in which
> imagination has the greatest role.
>
>   —Jean D'Alembert (1717–1783)

Imagination is an ambiguous occurrence that though sprouting beyond the exertions of thought and calculation, flawlessly flashes the most important of renderings in our endeavors into knowledge and creativity.

The feat of conceptualization is a natural force that is the crux in grasping the deep wonders of concrete existence, and it is the intensity of concept that is accountable in setting down the foundations of the two farsighted subjects: mathematics and theoretical physics. A flash of a concept is corollary to the installation of an unclouded field of vision, sojourned abaft the neural flaps—which in day-to-day language can be inferred as an instance of imagination. Accurately, this would be an engraving of the intuition beyond the whir of day-to-day miscellany. On linguistic grounds, "intuition," and not "conceptualization," is perhaps the most accurate term in describing the crisp cast beyond the bound of calculability.

It is the materialization of intuition that has served as an induction of anything that has emerged to be avant-garde in any area of humanistic endeavor. And although thinking betokens an exertive effort, the implantation of the intuition occurs only during the relaxed state when harmony settles under the act of conscious vigilance. That is to say that under the canopy of watchfulness the intuition peers out, cutting the haystack of entropies afloat at the subconscious level. The intuitions and thoughts that have changed the way we perceive the universe and ourselves, or that have given new meaning to doctrines and theories, struck during such composed states. This uninhibited state at the psychical level sums to a plenary vigilance of a cyclical duo of formative energies—enthalpy and entropy.

Let's first take in a little of how we see this duo in the order of the universe.

Equilibrating Enthalpy and Entropy of the Absolute Makeup

The scientific branch of thermodynamics—formally defined by British physicist W.T. Kelvin and derived from the works of many eminent physicists across the centuries, such as Joseph Black, Nicolas Carnot, Rudolf Clausius, James Maxwell, Ludwig Boltzmann, Max Planck, Josiah W. Gibbs, and, indeed, W. T. Kelvin—deals with understanding the energetics by the order of heat for a given physical process or attainment of a given physical outcome. For example, the binding of two molecules, say—a protein and a ligand—can occur only if the process is energetically favorable (i.e., if the product acquires a lower energy state). The free energy—formally known as Gibbs free energy (ΔG), after the American scientist J. W. Gibbs—in brief notes the potentiality of a process to occur, and the process is naturally propelled toward the posture that gains the lower energy. The reactants glide down to a more negative energetic state. And for a process that involves binding between the two molecules, the more negative the energy becomes, the tighter the binding, the steeper the gliding down path, and the higher the favorability of the progression.

A physical process in an enclosed conformity, or a creation of a physical frame, involves equilibration toward moving the energy level to its minimum; the lower the level into which the energy ripens, the stronger the structure of the resultants. The Gibbs free energy (ΔG) incorporates two contrasting exertive dispositions: enthalpy and entropy. Enthalpy, from the Greek *enthalpos*, "heating," states favorable interactions between the components, referring to the bonding between them, via, say, electrostatic forces. Entropy, the hint first given by Rudolf Clausius and later on resolved by Ludwig Boltzmann, conversely, points to the state of disorder, or complete freedom, that every component in that very system strives to take on. Enthalpy and entropy thus are qualities that counteract each other, both playing roles in the staging of the free energy for a given process. Gaining by enthalpy is concomitant to losing by entropy, and vice versa.

From the scene of the watcher (the same as the referred-to observer in the earlier chapters), which is recollected at the juncture of self-realization, the overarching sweep—the one of which the innermost ingredients are a part as well—collectively conforms to the same energetic mechanism that the dynamics of the universe integrates: the adjustment between enthalpy and entropy for any given activity. By its truly unembellished identity, the watcher would witness the potentialities of the eyewitness universe on par with the infrastructure of the inner plane; both add up to a display of a single uninterrupted holding, projecting coupled expressions of a single stroke (more on this later). In reference to the exclusivity of the pure observer, the structure of enthalpy flares as the mobilization of forms, while the spur of entropy would appear as hazy disorderliness, given every transition that the plenary latitude comes upon. And it is in reference to the pure observer only that the state of calmness sets in by the apperceptive dissociation from both the energy states—the orderly and the disorderly—where the worldly character itself becomes the observer, bolted next to the trimmings of the energetic etiquettes. It is in the tranquilly composed state, which engages the changing into the watcher beyond the bounds of the energetic setups, that the visions become clearer and the intuitions are manifested.

Thus, in the dynamic whirlpool of the full-scale mass–energy modulations, the launching of the intuition bespeaks an appearance that waves in consequence to the procurement of the role of watcher, as from the site of the watcher, the physical and the mental—be it a thought, an intuition, an idea, the scope of concrete three-dimensional landscape, or the state of worldly affairs—all would sojourn with unclouded edges, disengaged from the whimsical hues of entropy. It is in the state of watching that not only the code of enthalpy but also the rule of entropy would appear just and necessary in effectuating the uniform flow of space-time. The reasoning behind this has to do with the harmonious procession that occurs only by way of the counterbalancing actions of enthalpy and entropy, and the full view of enthalpy–entropy coupling would occur in the plane of the watcher, for the watcher exists beyond the all-embracing dynamic game in which the roots of the subconscious take a role as well. I will get back to this in successive chapters.

The coupling of order and disorder in the casement of realism is scientifically savored and mathematically seen. The beautiful order of chaos in the tone of mathematics can be relished in the title *Does God Play Dice?* by Ian Stewart.[1]

In the previous chapter, we came upon an explicit scheme that pitched an uninterrupted sweep symmetrically jelled with the plane of the steadfast watcher and recognized that the awareness of all of the elements of space-time in the totality would be uttered as a unitary universe, in the higher-dimensional field. A similar structural idea of self-containment emerges from the proposition "self-aware-universe"—the scheme descried owing to the assimilations from ancient philosophical and spiritual scriptures and their scientific reinforcements by the dispositions from fundamental and quantum mechanical physics.[2] The idea that the existence of space-time is tantamount to the universe being aware of the self—a self that encapsulates the matter and the mind impartially—is just from the perspective of the exhaustive awareness (being sensually and mentally aware of every beat that surges), from within the texture of space-time. However, the denouement in the voyage to converge into the hindmost ultimate would be the occurrence of the watcher unflappably witnessing the continuum of space-time, not from within but persisting at a distance, beyond the bounds of the ripples of the mass and energy.

Hence, in that sense, though the trait of awareness, or the consciousness, emblematizes a factor that is soundly interwoven in the tapestry of space-time, it is transparently different from the span of watcher, residing independent of the unfurling of the continuum.

It is not the equality between mental and physical but rather the equilibrium between the inner and tactile vistas that would sustain the throbs of this universe—as, ultimately, all aggregate into a single spread. In the affirmation of the ultimate identity as the bystander of space-time, the human-self eventually becomes the universe-self, in the flow of the collective mechanics (figures 7.2; 8.4). And in such a tactical layout, the human-self would be an illusionary manifestation in the dreamy state, whereas the universe-self would be a de facto manifestation in the awakened state. A self-realized being would acknowledge the unbound awareness, of the somatic terrain that culminates in the encompassment of the watcher, outlying the scope of materiality.

A barefaced query comes to pass. If the apical status of our own inhabitance is the unadulterated witnesser, why do we instinctively not know the spread of such a lofty realness? This acknowledgment beseeches the entanglements of the dream to be cracked—a maneuver that requires a disruption in the flow of the physical forces, the very extortions by which the across-the-board dream hangs on. And the ceaseless tide of physical forces, that though obliviously occurring in conjunction with the watcher, keeps the sense of the absolute state benumbed.

Utilizing leading-edge scientific understandings, here I will strive to dissect the role of dualism—the feature that permeates the universe in endless ways—toward gaining a perspective on how the rule of duality can explain the stamp of consciousness in the orderly design of space-time. And then we may be able to extrapolate the same codification to capture the print of parallel universes, and the eyes of consciousness that permeate them. The show of the duo comes in endless flavors, and we have seen many of them here. The pairing of enthalpy with entropy, space with movement, matter with its forces, and the full range of symmetries in the universal anatomy are all shades of twoness.

The String Theory Again!

Prior to the advent of the M-theory, brought upon by the works of physicists Edward Witten, Ashok Sen, Chris Hull, Paul Townsend, Michael Duff, John Schwarz, and many other leading scientists, the string theory came in five varieties (as I briefly noted in "The Quandary of Dimension" in chapter 6)—Type I, Type IIA, Type IIB, heterotic $SO(32)$, and heterotic $E_8 \times E_8$. These five theories—though they describe the union of the physical principles in much the same way, claiming that at the most fundamental level space-time continues by the vibrations of one-dimensional objects (the strings, as noted previously)—are built on different structural and theoretical frameworks; the distinctive methodological details vary. The launch of the M-theory, an advanced form of the string theory, stemming from the descriptions offered by Edward Witten, provided an overarching eleven-dimensional scheme that served to fuse the previous five different string theories into a single structural framework that involved a clarification by which the bothersome disparities between the earlier five string theories melted away.

Edward Witten identified that five different configurational frameworks are just five different ways of looking at the very same underlying physics, mediated by an impression that in the world of physics is known as duality.[3,4] The five forms of string theory that existed independent of each other fell into a single unanimous relevance simply by the realization of an underlying pattern, despite the outward façade of distinctiveness.

An Uninterrupted Show of Duality

The impression of duality, aside from being cast in the advanced theoretical descriptions and mathematical conceptualizations, in many different ways also imbues the fashion in which space-time itself is beheld. For instance, the existence of straight-out and hidden annexes, or the sway of matter and antimatter, from the perspective of the watcher, is an articulation of duality because ultimately they purport the two different modi operandi of the very same underlying occurrence. The dash of symmetry in between the mass and the energy, the positive and the negative charges, the mass and the gravity, the particle and the wave, the space and the time, the space and the movement, the contrasting emotions, and the supersymmetrical integrations that are recently staging the modern scientific views all emulate the decisive posture of duality. And in the arena of the mass–energy dynamism there runs an utmost case of duality—hammered between the image of the universe and the impressions of its parallel forms, executing evenness in the format of the superfold.

The anatomy of the mathematical manifold—a quintessence in understanding the makeup of truism—incorporates an orderliness, addressed as Poincaré duality, originally captured by French mathematician and physicist Henri Poincaré (in 1893). His argument pointed out that in specific pairs the homology and cohomology groups of the manifold—a closed (compact and boundaryless) algebraic structure—transpire isomorphically (i.e., they are structurally equivalent, in a nutshell stating that these two geometrical conformations are just two different manners of seeing the exact same fold). The phrasing of Poincaré duality provides ways to comprehend the complexities that emerge in modern forms of the string theory and help capture the sketch of the manifold universe, the materialization of which is unfailingly concluded in quantum mechanics.

The two isomorphic groups are equated as follows:

$$H^i(M) \cong H_{n-i}(M)$$

In the above equation, M is a manifold; ith and $(n - i)$th are the cohomology and the homology groups, respectively; and $\cong$ denotes isomorphism.

The plot of duality is unconditionally felt in the manner in which space-time flows and crystallizes. The order of the universe and the distinctiveness of the accompanied parallel universes that stem from the musings of the quantum mechanical projections must be caught, afresh, in conjunction with the tenancy of the conscious realm. The picture of parallel universes that emerges in physics and mathematics can be evidently viewed to be definite, once more, by welding the elements of subconscious that stay permeated by the continuum of the watcher into the scene of reality. And with the embracement of the watcher we have an order of a paramount duality that we will come to later on.

Putting it the other way, in realizing the schematic of the side-by-side ceaselessness of parallel universes, the authenticity of the watcher, latched alongside the swarm of the quantitative folds, becomes all the more conspicuous. Let us try to scan this.

The Culmination of Parallel Universes by the Scene of Duality

Every universe palpitates through the assortment of the self-held elements. Quantum mechanical interpretations profess the existence of many worlds (the multiverse or parallel universes) by affirming the conduct of universal wave function—shown by American physicist Hugh Everett III—which accommodates all of the probable histories and futures, streaked side by side, instead of a single chosen one. And in reference to Heisenberg's uncertainty principle, which states that certain pairs of physical properties, such as position and momentum, cannot be measured simultaneously with equal precision, the existence of multiverse points to the fact that, as indicated again by the works of Hugh Everett (in 1957), such uncertainty in the admix of all the probabilities implies that each of the histories that occurs in reference to an observer is held sequestered, for its observer goes on oblivious to the whispering of the other parallel occurrences.

Thus, the theoretical emblem of the multiverse in reference to the observer (occurring for each probability), transpired in quantum studies, again emulates a scene of duality. The observer, although itself residing beyond the hums of the elements, is projected multifariously, and once we are able to pin down the accurate positioning of the observer—the observer that observes the subliminal functionalities as well—we may be able to further applaud the authenticity of the quantum mechanical expositions in discovering the tides of the parallel universes.

Here the stunt of duality can be appreciated at the most apical view. In the streaming of the multiverse, the coming about of a universe would be an outcome of distinctive watching—of only a fraction of the functionalities, out of a large pool of the universal set of functionalities that exists solitarily. (We are elaborating this next.) And in the establishment of the parallel worlds, neither the flurries of elemental activity nor the stroke of the observer reside in dualities. The scene of duality is evinced in the wake of the observances of the same functionalities multiple times. Let us try to delineate this.

Singling Out the Core of Nonduality from the Grand Frame of Duality

The coming about of parallel universes can be seen as an outcome of sharing the elements that exist between them. I laid out in the preceding chapter how the flow of an exhaustive mass–energy is akin to the watcher being aware of them, autonomously mirroring an implant of an individual universe. Therefore, toward figuring out the landing of the parallel universes, instead of the scenario that each universe bears its own set of materiality, a schema that there strings a single universal set of elements—where the parallelism springs up because the watcher, although manifesting independent of the billows of space-time, turns up to be identified with a section of that universal set of elements—not only underlines the state of nonduality infused into space-time but also brings out the steadfast seal of nonduality in the span of the watcher.

The force of nonduality emerges as the most integral thrust in the flow of the grand unified field, for reasons I will explain soon.

The kicking in of the division (or duality) in absolute landscape is concurrent with the effectuation of "states" referring to "observing or being aware of" only some of the elements out of the large generic set. Say there is a generic set of a thousand elements. A unit universe can come to exist if the observer or watcher is aware of only some—say, one hundred—elements of the generic share. Those one hundred elements would be warped by the forces of physics in association with the continuum of the watcher, spewing a solitary self-subsumed universe. The observance of another set of one hundred elements, that has overlapping elements with the previous universe, would spit out yet another universe side by side with the previous universe, and so on (figure 9).

The scheme of nonduality, not just in the state of the watcher to which the seers submit but also in the state of space-time, is not only a dire necessity in explaining the pounding of the parallel universes; it also brings forth a picture of how the energy itself is manifested optimally in the grand weave. A non-multifarious mass-energy pool (in the mold of the parallel display) (1) is energetically economical, for all laws and forces are implemented through a singular set of mass and energy, and (2) by the laws of thermodynamics would allow the attainment of a global minimum by a single set of enthalpies. The feature of nonduality, thus, is rather an automatic development, like any other property that the ultimate design staunchly displays.

A point in time in parallel manifestations

Figure 9. Schematic behind the induction of parallel universes. The universal set of elements is shown as an assemblage of filled circles. A smaller set of elements (shown with oval boundaries) streaming in coherence with the empty space (the watcher) would constitute a solitary warped universe. The elements are shared between the aligned universes. Gradient filled circles (in the center) signify the most common elements of the pool.

Foremost, the hallmark of nonduality appears to be acute for the exhaustively interactive continuance of space-time.

The grid of the all-embracing generic set not only incorporates all of the physical functionalities but subsumes emotive and psychological gadgets as well, exhaustively. The imprint of a single universal set, as opposed to a motley of segregated ones, is foremost an essence of the functions and operations that occur in the vital universe, and so in the shadows of parallel universes. Light examples of "shared elements" can be taken from the domain of conscious embodiments. For instance, a pair of human beings, human A and human B, who interact with each other do not epitomize the modicums of two segregate elements but rather would equate to a strand of a solitary element—the interaction between A and B—which is a functionality. A human C and human D who join in a classroom, where human A is the teacher and human B is the student, again exemplify an instance of a single element—a solitary goal with two roles. The ongoing operations are possible only when there is a continual landscape of a single overarching universal set because elements are needed to be shared.

The setup of parallel universes, like so, does not imply the presence of several copies of the cosmic universe into which each of us would bump. In the scheme where the absolute function is the watcher, every human entity entails the watchfulness of the functionalities (palpable and internal) that overlap with the functionalities that any other human being comes across. Thus, at the psychical level as well, an element (for instance, an emotive bearing) that is shared between humans A, B, and C will be detected as a show of three copies—in the quantum mechanical descriptions—while in actuality that element would be present in singleness between the cast of three humans: in relativity, at a given point in time.

The Sweep of Consciousness in Parallel Arrangement

With the ultimate coherence being the watcher, and not the functionality, the being-self evolves to the stature of the cosmic-self—the space-time warp by the function of the pure observer (figure 8.4). The multifarious three-dimensional universes, that include the pleats of the psychological reach as well, therefore converge to the flaps of multifarious selves. And in this unconditional landscape of space-time that motions with the scan of the observer, the skeleton of the parallel universes becomes analogous to the verity of the parallel selves, or the observers. There are common as well as uncommon elements between any two "universal selves" (figure 9), setting forth a slight asymmetry in the panorama of space-time, a characteristic that is essential to the flow of time. The sharing of the elements between the two universal selves in the web of physicality would correlate to a situation where the independent beings, when gazing at a solitary object (or together experiencing an activity) neatly positioned in the mesh of the space-time continuum, spatially stay close in vicinity. The instance of the eye catching a sight corresponds to the observer being aware of its universal and mental sides, for every given point of time—as that is all there is to be aware of for that juncture.

By the same token, the functionalities of the psychical sphere, such as career choices that we make, subjects that we enjoy, the way we like to entertain ourselves, the places we choose to live in and visit, our emotional states, and so on, would all be part of the grand pool of the nondual set. Thus, it can be appreciated that the denouement in the game of dynamism is not about how the two conscious embodiments are lodged but rather that the observers are observing the elements that are shared between them, a situation that would effectuate the matrix of parallel universes in the quantum arena.

The notion of shared elements can be grasped more clearly on sentimental grounds. Between the sentient beings, the higher the number of common elements, the more comfortable the psychological states (in being together). And a lesser number of shared elements would imply lesser (and weaker) enthalpies, consequently radiating lesser psychological comfort (in being together) for a given point in time. Indeed, because the whole stage of space-time constantly shifts from one strike of time to the next, the shared elements between conscious embodiments also rearrange wholly, including the addition of new elements and the subtraction of some of the old ones, for every space-time beat.

No History Left Behind

Quantum theories suggest that the universe ripens via all possible histories.[5] The universe in its current state did not come to exist following a single defined path; rather, every probable path has been followed. Thus space-time doesn't have just a single history; it carries all the statistically possible histories. Tracery of the path that a particular moving point (quantum) particle has followed through time is referred to as the particle's worldline—a term coined by German mathematician Hermann Minkowski (1864–1909), who pioneered the idea of time being a dimension in the fold of reality, which was later advocated in Einstein's work. The worldline, thus, elucidates hitherto the history of that particle. The blueprint of a vibrating string's history, thus, will be seen as a surface of a two-dimensional manifold, termed as a worldsheet—an expression coined by Leonard Susskind, a physicist at Stanford University, who is regarded as one of the leading authorities in the establishment of the string theory. And because the string theory seeks to accommodate all possible probabilities of occurrences, the landscape that emanates from this theory axiomatically embeds all the traceries of history that can possibly emerge.

The sheetlike surface that is emerging in the progressive formulations, referred to as a membrane or brane, which incorporates higher-dimensional fields of interactivity, further schematically accommodates all possible undulations, with a full range of energy upsurges and downturns, and even allows local tearing in the structure of space-time—a phenomenon that has been seen to occur in an overall perturbative scheme.[6] Every surfaced ripple in this sheet announces the emergence of a mass-energy form followed by its annihilation, in conjunction with the antiparticle. Therefore, the membrane-like quantum appearance prevails through all the histories, denoting the past lived, up until the current point of time, by those mass-energy forms that constitute the makeup of the membrane. History—a path by which a mass-energy manifestation has come to exist—is not an averaged or a universal path; it is the path that was taken in reference to an observer that is latched alongside the course of that history. The history occurs in reference to the observer.

Thus, every stamp of history (a worldline, a worldsheet, or a life course), though intricately enmeshed in the membrane tapestry, would transpire only in relation to the observer. The ultimate eventuality of any journeyed course is the state of the observer, which, though it persists beyond patterns, the eventuation of history stems only by the blend between the mold and the watcher. Therefore, for the higher-dimensional brane, the number of histories would relate to the score of projected observers that encapsulate selective sets of elements, casting the pattern of parallel universes, shifting for every beat of time.

In reference to human history, the life span, the observer encases the alliance of the cerebral and worldly entities through the course of life, subsuming both space and time components—spewing a framework of the universal self, and streaking a history in the arena of the quantum undulations. Therefore, the conscious sweep would entail assuming a mass-energy form for every given point in time throughout the course, where every stamp of mass-energy at the ultimate level is the observer of the functionalities in that lifetime.

The constant jitters and the existence of all possible histories that we acknowledge at the quantum level correlate to the actuality of the parallel universes that dwell in reference to the watcher when we account the inner bearing in the picture of the final weave. When we take into account the beats of all mass-energy manifestations under the influence of the physical forces, constant jitters and existence of all possible histories in the sweep of the quantum brane become a consequence.

The Case of Missing Undulations in the Cosmic Uniformity

The undulations of the quantum world, or the existence of multifarious universes, cannot be acknowledged cosmically because the cosmic encountering stems via the five senses for every given point in time, in respect to the observer that beholds the course of the particular lifetime, and thus only a solitary worldline can be witnessed—leaving out the other histories, the same way Hugh Everett III, based on quantum mechanical studies, highlighted that the observer of one universe stays totally unaware of the other universes that flap in alignment.

The empty set (space, $\{\varnothing\}$) plays the same role in the existence of the multiverse as it does in the existence of the solo universe—its intrinsic latching in the development of the boundaryless continuum. The numerations erect only in conjunction with the empty space that lies in the background. Every instance of count conceals the empty space, by dint of a boundary that the count itself casts. With the skewed sense of selfdom—that is, being identified with mass rather than the empty space—the illusionary boundaries appear evident, resulting in the flare of paradoxes and dilemmas, and foremost, we stay put in the state of a grandiose dream.

Multiplying a number with 0 (or an empty space) results in 0. However, it is not possible to divide a digit by 0; a routine simply cannot be devised. Zero, echoing the essence of the empty space, is a state that sweeps beyond splitting, and the routine of dividing causes the denominator to vanish. A hypothetical attempt to divide by 0 would challenge the removal of the numerator beforehand— plainly because the materialization of any numeral occurs only in alliance with the empty space ($\{\emptyset\}$).

CHAPTER 10

P

The Fractional Universe: The Blanket of the Subconscious

The Behaviors of Particle and Wave Espied through Two Separate Windows

The Duality of Bidirectionality
in the Hammer of Reality

The ladder of arithmetical exponentiation reflects bidirectionally. For example, $4^2 = 16$ is an exponentiation of 4; $4^{-2} = 0.0625$ is also an exponentiation of 4, only in an inverse direction. Exponentiation in reverse exhibits nothing but power fragmenting the state of *one*; 4^2 is 4, 4 times (16), while the routine 4^{-2} doesn't reflect ciphering a number that when self-multiplied begets 4—it inquires for a number that, when self-multiplied, breeds into 1. As it goes, the expression 4^1 declares 4, while 4^{-1}, which betokens fractioning the 1 by 4, giving a chunk of 0.25. Likewise, 4^{-2} is again fractioning 1, two times successively, by fours, resulting in 0.0625 (0.25/4). By the pure arithmetical logic, the slice 0.0625 appears to be a far smaller expression compared to the totality of 16, but just in reference to the functionality of exponentiation, the results of 4^2 and 4^{-2} ascribe identical displays. How many unit elements do we have as a result? Both 4^2 and 4^{-2} result in the appearance of 16 elements. It is the appearance of 16 units that emerges in the routine of 4^{-2} (0.0625 presented 16 times) as well.

The arithmetical premise of exponentiation is an articulation of a bidirectional ladder that expands from the station of a unit—transparently of 1. For example, in the exponentiation where the base is 2, 2 epitomizes that unit, which is rooted at the origin of the ladder, and here "2" would be exhibited as the main whole, signaling the exactitude of a unit. Likewise, if the base happens to be 3, then "3" stands for the integrated totality that exhibits the foundation of the progression (figures 10.1, 10.2). In climbing up the ladder, the entirety multiplies itself, while descending exerts partitioning of the same collectivity, generating an identical number of bits on both sides for every consecutive rung. The exponentiation is an induction of branching for a unit. This is precisely why the exponentiation by

0 (for any base) uniquely leads to 1, not to 0 or any other numeral, irrespective of the base size; 4^0 is 1; 1^0 is also 1. The exponent 0 clues us in that the unit-whole did not submit to fragmentation or stayed as it was; no branching occurred. With the exponent 0, the attribute of multifariousness is lost, regardless of the order of the base. In like manner, negative exponentiation by 0 (for any number), X^{-0}, also comes to be 1 for the same reason: the totality did not fraction, and it stayed the same way. To some, these jottings on the procedure of exponentiation may appear very basic. However, in view of the glistering model that this simplistic enunciation displays in mimicking the truss of reality, it would be a dire oversight not to reflect over the bearing of exponentiation contextually.

In this bidirectional ladder, the world on one side of the entirety would be a look-alike of the world on the other side, and from the perspective of the waving singletons, both the worlds would be on par for the germaneness conveyed by the texture that they put on. The contour view of the pattern will show a symmetrical setup of expanding branches on each side of the plenary unit. The entirety that constitutes the center projects in both directions (figures 10.2 a, 10.2 b), erecting an image of two facets cruising in concurrence.

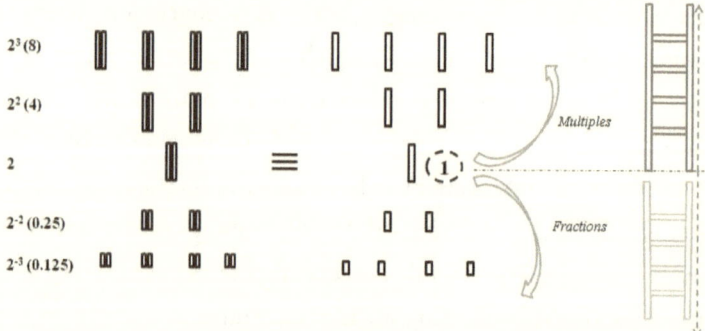

Figure 10.1. Exponentiation is reflected bidirectionally. The numbers of unit elements on the both sides are identical for every consecutive rung. For example, the last gradation on each side has four unit elements in a series where 2 represents the unit.

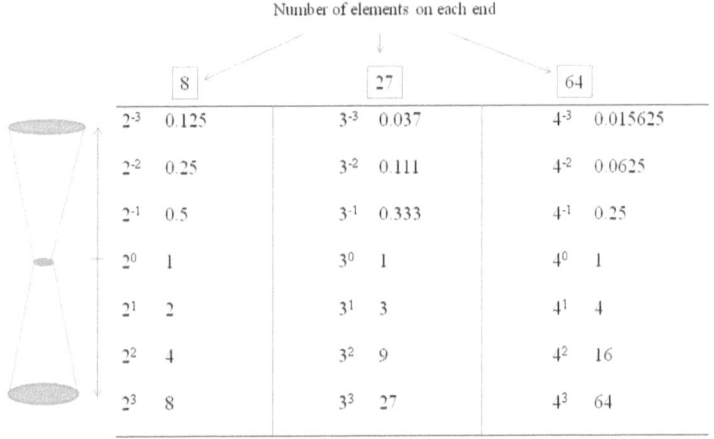

Number of elements on each end

8		27		64	
2^{-3}	0.125	3^{-3}	0.037	4^{-3}	0.015625
2^{-2}	0.25	3^{-2}	0.111	4^{-2}	0.0625
2^{-1}	0.5	3^{-1}	0.333	4^{-1}	0.25
2^{0}	1	3^{0}	1	4^{0}	1
2^{1}	2	3^{1}	3	4^{1}	4
2^{2}	4	3^{2}	9	4^{2}	16
2^{3}	8	3^{3}	27	4^{3}	64

Figure 10.2 a: Irrespective to the base, the exponentiation is a bidirectional projection of a unit that intimates "1." The negative and positive exponents dispense identical number of elements, over fragmentations and amplifications, respectively.

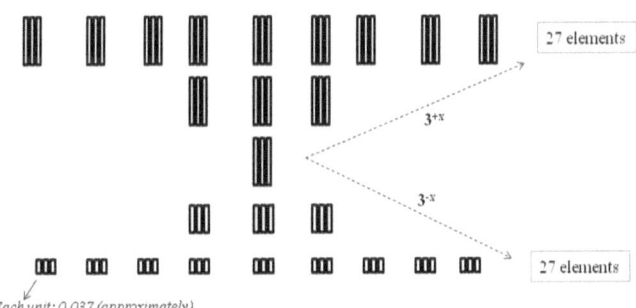

27 elements

3^{+x}

3^{-x}

27 elements

Each unit: 0.037 (approximately)

Figure 10.2 b: Irrespective to the base, the exponentiation is the bi-directional mushrooming of a unit. 3^{3} generate 27 elements. 3^{-3} also produce 27 elements, weighing 0.037 each, that are generated over fragmentations.

Positioning the Five Senses in the Grid of Space-Time

In reference to the unbroken reach of the watcher, the clasp of five senses places the world on one facet and adumbrates the likeness on the other (the layout we have contemplated in the previous chapter). The sketch of likeness shouldn't be taken as literal here, and rather implies a symmetrical setup in which the psychological elements also play out. In the context of the bidirectional exponentiation, the fractionations would tally the veiled phase of the universe. The outwardly directed sensory seizing of space-time brings about the casts of the inner veiled and the outer exposed. From the viewpoint of the watcher, both facets remain part of the same stiffness.

Even more so, the polar behavior in exponentiation crisply matches the appreciated actuality of nonduality, of the mass-energy as well as the watcher, in the amassing of the parallel universes. In the vitality of the multiverse, the coming about of branching alludes to the multiple observances of the exact same unit, a single blanket of the universal set, which has subliminal elements tailored into it. The point of bringing the abridgment of the counteractive exponentiation has as much to do with calling attention to the beat of symmetry that occurs between the imaginative nature and the ostensible universe as it has to do with pointing up the juncture of the deflection of two, or more, from the conduct of one—whether we discuss the inhabitance of a single set of universalities or the continuance of the watcher beyond it.

The recondite subconscious of the inner phase can be seen in very basic terms as a sum total of all neural experiences and emotional effluxes, and, in all possibilities, some deceptive commixes of the two.

The administration of duality under the grip of misjudged identity—of the matter and not the observer—unfolds to reflect a cascade of two distinctive autarchical existences. In the sound of mathematics, this duality packs such an arrangement that if the branching of entities on one side occurs by the means of multiplication, then the other side must manifest antagonistically—via fractionation. There is no alternative way to articulate both equivalency and distinctiveness of the planes. In the game of time and energy, the mental outlooks and the worldly gestures not only are equally important but also coexist.

In view of the conformity between the inmost and the exterior, the subconscious kindles with the worldly front in the execution of a moment. By the same token, the encounters of the world would be an essential causation in drawing out the elements of the subconscious mind, and the death would be the end of both—that which existed in connectivity for the observer.

Why the Duality of Wave and Particle?

The argument of wave–particle duality calls attention to the potentiality of a coupling—wave with particle—in the comportment of reality. The theory is pillared on a composite of mutually contradictory advocacies on the nature of light, proffered by the celebrated physicists Christiaan Huygens (that light signifies waves) and Isaac Newton (that light stands for particles [from the seventeenth century]). The accuracy of this argument of polarity has been unfailingly seen not just in quantum interpretations but also in the dispositions of the compounded states of atoms and molecules. The Copenhagen interpretation—that the act of measurement causes the probability to shift toward assuming one of the states, the wave or the particle—highlights the beat of the wave–particle complementarity, which can be seen either as wave or particle but not both simultaneously. And, although stemming from different chains of thought, the particle–wave complementarity stands to be analogous to the equivalence between mass and energy that Einstein pointed out—that mass and energy are one and the same collectivity woven in the relative décor of the cosmos.

The very attempt to determine whether the actuality is a wave or a particle instills a bias of what must hatch. And yet it is not possible to know the nature of the form independent of the measurement. The dab of this Copenhagen's complementarity is prevalent in every aspect by which the universe goes on. The conglomeration is an essence for the time to flow because this flow necessitates a continual change—a steady succession of transformations—and it is in the state of comutuality that the perpetual state of changeovers can take place. We may be able to acknowledge this in just a bit.

No matter how perplexed or inexplicable we feel about the entity mind, we all know with no skepticism that the surges it emanates are no less real than the arena we lay eyes on. If this is to be so, the aspects of mind must, in some way, reflect a physical form. Energy is the cast of physics as much as the mass is; anything that exerts and pounds is, on the universal stage, a function of reality. And the journey that suggests metamorphosis at every instant—coalesced with the continuance of ticking—insinuates the shifting in the psychical texture for every moment as well, in tune with the ongoing transformations in the structure of the grand regularity.

The complementarity in the nature of reality is required because the progression and transformation are synonymous, and the one that transforms is the particle; the particle transforms in the progression of the wave. Recall the transformation that manifests after the causative annihilation occurring between the matter and the antimatter in an all-out interactive field (from chapter 5); the world–line (or the worldsheet) that is evoked in this field of everything refers to both particle and wave.

In reference to the all-encompassing view of the observer, the body can be regarded as emblematizing that particle, much like an eigenstate (discussed in chapter 1, under "Procuring a Panoptic Window"), where the full-length life course would constitute a worldsheet. The blending is essential to space-time because the flow of reality entails the embodiment as much as it calls for the movement, through each and every constituent fiber. The entities across the board must imperatively materialize as they must also obligatorily sustain mobility, so much so that every material must incorporate dynamism and every movement must beckon an entity. Thus, the formation and the progression, in the rule of reality, must remain utterly indistinguishable. The transformation that is elemental to the carving of the world–line occurs through the uncompromising changes that are collectively manifested in the unifying arena, shaped by the admix of the subconscious and the universe—where both constituents together compose a trail that is witnessed as the world–line (or the life course) by the dwelling of the watcher. For the unfettered watcher, all hovers in the cast of singularity, which includes the elements of the subconscious in the panorama of the rhythmic enterprise.

In the unifying scheme that accommodates the consciousness, and so the subliminal, the materiality of wave and particle becomes the profiles captured from two different windows.

Peephole on the Behaviors of Wave and Particle

The valuable insight of Heisenberg's uncertainty—we cannot estimate both the position and the momentum at the same time and with equal exactitude—asserts that in the pairing of the physical peculiarities, only one of the counterparts is reflected. The act of measurement dictates which one of the two gets procured. The equivalence of the two peculiarities doesn't imply the cast of the two states in an equal fifty-fifty way, the way one could assume the effectuated complementarity to be, but rather reason that the two states chime with two dissimilar facets of articulations. And thus, in the panorama of the systematic unison, it is not that the act of measurement dictates what the system evolves into, but rather it is the case that which of the two peculiarities shows up depends on the choice of the window through which we peer. There abide two separate viewing apertures: one through which the particle is seen, and the other through which is beheld the wave.

In fact, we will come upon the appreciation that out of the two views, one streams "discrete" embodiments, and the other encompasses both "discrete" and "wave" components.

Both of the feats, the particle manifestation and sinuosity, play equitable roles—both are accountable in the enforcement of every mark of time. The frame of duality is actualized for the planes of these two windows reside distinctively; the two windows themselves simultaneously manifest for every speck of the moment. This translation, rooted in the inferential clues and the picturesque view of the watcher, suggests that the intonation of duality, or the continuum of complementarity, that is reflected for every speck of the moment is in itself a reflection of the continuance of these two gazing windows.

The commixing of the bodily and the subliminal planes, for every strike of time, concocts an entity. The feature of physique chronicles the sinuosity of inner span the same way the mass recites the energy, the space reports the time, and the particle recounts the wave. Although both the pairing aspects are witnessed in the field of the watcher, only one can be witnessed by the force of the electromagnetism. The lens of the watcher and the one of the eye are the two disconnected fenestrations in the workings of space-time. A manifestation of whether it is point or wave, bodily or sweeping inner, or clean-cut or inexact depends on the location of the window.

The outer–inner continuum thus instills an element, for every beat, that exists through the fundamental forces, under the impassive witness of the onlooker.

Throwing All into the Possibility of a Grandest Dream: A Fractional Universe

While asleep, we repeatedly gaze into the screenplay that is themed around the subconscious disposition—a screenplay as evident as the piquant show that streams on the screen of life. Awakened to the staunch play of the world, we realize that what was perceived as a solidity of body is, in the real world, the cast of subconscious. Fallen into the bubbles of the physical forces, the rolling out doesn't make the phenomena of functionalities any less cogent, for the interactive spree that takes over the subconscious play, where the live characters collectively merge into a single real-life movie.

And in such a grandiose setup it is the presence of the watcher that gets eclipsed, owing to its own lack of recognition of, again, the watcher—its sheer identity.

Beyond the neural encountering of the discrete universe, all of the elements come to be the strands of the inner weave, which will include the "physique," as well as the world that connects to it via the textural forces. Two parallel universes kindle through the overlapping elements of the deep field, where each of the universes remains cocooned by the act of witnessing. A dab of solo universe in itself is a fractional existence because a universe can have only a part of the whole set, or it exists only in concurrence—through the unabbreviated field, an entirety that symbolizes 1. The totality can be effectuated only by the synchronicity of the parallel universes in the plane of the fractional ones. The partiality of the lone universe would grow to 1 in the higher field of the parallel universes.

About a year ago, I chanced upon a fictitious, but nonetheless thought-provoking, flick—*Inception*. The theme of the movie revolved around the idea of breaking into someone else's subconscious by the intervention of a shared intravenous drug that would sedate the participants into a dreamy state in which the subconscious elements of all the members would be woven into a single network of occurrence. The punch line, however, was that this ploy was set up in the needful longing to extract conspiratorial information from the subconscious of an individual who otherwise, in an awakened state, would keep the information secret.

If one could prudently travel through the vast basin of the subconscious, the inner personality traits would no longer be hidden. The most gripping angle of the story was that the lead character attentively kept aware in deep states of his own dreams, or in the stage of the connected dream that was shared; that is, he consciously knew that he was dreaming—and therefore, on the shared stage, he carried through according to the plan that he had designed when he was awake.

The story line is intriguing, with a tinge of reality dissolved in it, mainly for two reasons. First, it sees the possibility of keeping aware when dreaming, consciously knowing that one is asleep as one catches the illusionary play. And, second, it calls for a mastery of vigilance, demonstrated by the lead character, of knowing one's own subconscious in totality. The lead character himself had a subconscious of which he was fully aware. Therefore, in the state of shared dreaming he not only knew that he was dreaming but he also kept fully aware of the cook-ups of his own mental arena, thereby bypassing any possibility of fabrication, emanating within him, that could deter him from the elected role that he needed to play; a cooked-up feeling could interrupt the focus that he needed to exercise in order to achieve the preset goal.

Although the reasoning is not explicitly charted in the story line of the movie, the full awareness of the subconscious that the lead character possessed was indeed the most elemental necessity in the execution of his role; the usual "unknown," the deep crannies of the subconscious, was fully familiar to him.

Knowing the subliminal in totality would mean being aware of all of the functionalities that float in us, not just the ones that peep out counteractively with the worldly elements in nailing the time. Upon the acknowledgment of all the elements, at any point in time, the essence of the purest appellation spontaneously unseals. One who recognizes all of the elements cannot *be* the element. And in the state of all-out awareness, the utmost uniqueness becomes the one who witnesses the full-blown engraving. From the vantage point of the gaze that captures the full-scale knit, all-that-there-is reduces to the arena of dream—in sleep as well as when awake—operating according to the laws of nature. The conduct of watcher is the pose of steadfast witnessing that spotlessly goes on through the two junctures of life: asleep and awake.

Even in the fictional plot of the movie, the main character can pull it off only if he gains a state of the watcher; or, in other words, if he mindfully spots himself in the continuum of the watcher. In such a turning point, his character-self reduces to a projection that is stringing in an all-interactive field of forces. The continuum of the watcher being the utmost of all, we all, knowingly and unknowingly, hover as the participants of a schematic that hems the traces of our own characters. The dream being the only arena, the credible distinction between being asleep and being awake would be akin to the difference between the solitary universe and its parallel forms. While we sleep, we aloofly deal with our own subconscious, but when awake, our subconscious becomes manifest in the overlapping mode, materialized by the pitching in of the other, parallel, observers and the associated elements.

CHAPTER 11

Perpetually Expediting Cosmos

Metaphorical Journey in the Field of Gravity

Probing the Truth of Movement in the Circuit of Space-Time That Is Compact and Closed

Tapping into the Exactness of Movement

No matter with what agility Achilles runs, he won't be able to surpass the tortoise, granted that the tortoise gets a head start in the race. Achilles will always lag behind, for he would need to pass through infinite footsteps that exist between any two points on the race track just to traverse even the smallest possible distance. Zeno's paradox stands for double trouble—conveying that the object is moving and not moving, both at the same time.

Attempts to delineate movement in the fabric of a continuum would surely lead to the flaring up of convoluted paradoxes, for the movement is spun into the fabric and isn't an autonomous exertion. The existence of two points in space in itself indicates the instance of progression (the way we discussed under "Essentiality of Annihilation in the Pulsation of Reality" in chapter 5). Thus, in the crystallization of time, what illustrates as an act of displacement in the tailoring of the continuum is the manifestation of the sequential lineup of "snapshots." And the tangled oddity of cutting across infinite steps eases in appreciation of the underpinning that what appears to be the stance of flow in the gist of it is a sequential succession that sets up under the witnessing of the watcher. The ploy is brought off the same way as the devising of an animated movie can be achieved by welding the still pictorials into a sequential arrangement.

The snapshots of the actual movement wouldn't just be the pictures perceived but would also include the "eye" by which the picture is noticed. No matter how many gradations there are on the racetrack, they would correspond to the still prints that are sequentially arranged to characterize a track—the course that can be seen as a materialized instance in the view of the watcher.

And from the stage of the watcher the movement correlates to the manifestation of the track, while the coming about of a snapshot corresponds to the stamp of a single phenomenon encircling both inner and outer extensions. Thus, in this alluring rendering, the paradox would not exist because Achilles does not run, only manifests; or, in the purest sense, being the watcher attests only to the race. The quandary sets in as soon as Achilles wanders off in the act of counting. This all might seem a little mystical, but you may be able to appreciate the snapshot scheme of reality after seeing some of the cosmological observations discussed in the coming sections.

From within the textural solidity of space-time, the act of progression or movement is the tone of existence. From subatomic particles to palpable objects to earthly beings to the revolving solar system to the whirling galaxies to the constantly expanding universe to the waxing of the intellect and the heightening of the instinctual drive to learn and employ, we cannot furnish a single example of occurrence that sustains but does not maneuver. The objects that appear outright fixed, too, are invariably bustling, including us ourselves, in the deep field of the cosmos. And it doesn't stop there. The vitality of dynamism itself is further reinforced for every given moment, seen as a constant acceleration, apparently in all the facets of reality.

Ballooning Cosmos: Restating Gravity

In the modern era of cosmological undertakings, it has been realized, quite unexpectedly and against the common scientific purviews, not only that the universe is expanding but also that the expansion itself is being continuously expedited. With the launch of the Hubble Space Telescope (in 1990)—named after an American astronomer, Edwin Hubble—which orbits outside earth's atmosphere and consequently captures extremely highly resolved celestial images, it became possible to access the ultra-deep fields of the cosmos. One of the major breakthroughs since the initialization of this device was the validated affirmation that the expansion in the universe is accelerating at an electrifying rate. The inputs captured by the Hubble telescope led to the awareness that in the deep field of the flying swarm of galaxies, not only are the galaxies drifting away from each other but also the farther out a galaxy journeys, the higher its speed of recession. The rate of the expansion of the universe is measured in terms of a parameter known as the Hubble constant. And although the proof that the universe is ballooning wasn't explicitly concrete prior to the advent of the Hubble telescope, the intellection of the continual enlargement and speeding of the universe had been a part of modern scientific assimilations nonetheless. The forceful extension in the vista of the universe was greatly recognized and mused over even before the arrival of the Hubble telescope's direct input.

The "metric expansion of space" is an astrophysical diction that addresses the increase in the space between two telescopic objects with the passage of time. Thus, instead of the belief that the celestial entities are rushing off from each other, the metric expansion of space conveys that it is the interspatial region between celestial entities that is growing, instead of the objects themselves traveling anywhere. Although such theorization holds at extremely large scales, such as when addressing galaxy clusters or the known universe as a whole, the astronomical bodies within the smaller ranges—which have larger in-built gravitational pulls—also have been chewed over for the departures between them. The conviction that the gravitational pull will eventually, in due time, cause all matter to crumble and recollapse—in the case in which the energy density crosses a certain threshold, reversing the expansion, ultimately concluding in a crunched singularity—began to wane in the wake of the great many evidences that show that despite the presence of the unifying gravity, the universe is continually not only just blowing but also speeding up, apparently for every time point.

Cosmological observations unwaveringly show a presence of an invisible or hidden, commanding nonetheless, scope, justly addressed as dark—dark matter and dark energy. Estimates suggest that the universe is composed mainly of this dark substance; the dark energy comprises 73 percent and dark matter 23 percent. All that we directly discern, including us, falls within the remaining tiny 4 percent of the entire universal material. It is conjectured that it is the dwelling of dark energy that ticks up the exponential growth in the spatial extent—the bridging that earned Adam Reis, at the Space Science Telescope Institute (a NASA center for Hubble Telescope operations and pertinent research); Saul Perlmutter, at the University of California; and Brian P. Schmidt, at the Australian National University, the Nobel Prize in Physics in 2011. Although the strength of these findings mainly stemmed from intergalactic surveys, for our own curiosity, an associated observation revealed that our own solar system moved about one hundred thousand miles per hour faster than expected.

The presence of dark matter was postulated to justify indirect observations according to which the amount of cosmic matter should be far more than what the "luminous" cosmos indicates. And while dark matter is seen to impart an invisible force holding together the galactic structures—a force akin to gravity—the dark energy, contrariwise, has been conjectured to be the force that led to the expansion of intergalactic regions, stupendously ripping the galaxies away from each other. Thus, toward tailoring the fact of hastening recession against the attractive pull that gravity enforces, it was suggested that it is the act of dark energy that unfolds contrariwise to the influence of gravity.

Popular science articles by Michael D. Lemonick give a swift account on the physicists' impression of dark matter and energy.[1] A simple and detailed account of this mysterious dark content can be found in the title *The 4% Universe*, by Richard Panek.[2]

An Integrated Hastening

The characteristic of expedited progression is signaled not only by the cluster of galaxies and the ballooning of the universe but also by the quickening in the breath of the intellectual capacities, in any area of human endeavor. It is not just that the rhythm of clean-cut commensurate dynamism propels the universe but also that the deep-seated dynamism in itself is aggrandized for every successive moment, through the horde of dispositions that texturize the supreme field. And in all of these constitutional augmentations, the most noteworthy pitch, perhaps, is the whopping expansion of the cerebral attributes. Whether we address the boom in technological advancements and scientific eruditions or the elevated awareness of psyche and emotions, the inner attributes too seem to be approaching a logarithmic phase, the same kind of hastening that we see in everything else around us.

An American astronomer, Milton La Salle, was the first to make the coincidental discovery that the galaxies not only recede relative to each other but also that the farther they fly into the abysmal, the more speed they acquire. He noted that a galaxy that receded with a speed of about 8,899,600 miles per hour in 1928 departed with a speed of 89,474,400 miles per hour in 1936; the speed increased about ten times over the course of eight years. The Hubble observations indicate that galaxies that are gliding billions of light-years away are withdrawing from us with hustling flights, with speeds reaching as high as 99 percent of the speed of light (i.e., close to 186,000 miles per second). In the blanket of the cosmos, the rate of dynamism, thus, is perpetually augmented.

A popular science title, *The Accelerating Universe*, by Mario Livio, an astrophysicist at the Space Telescope Science Institute, is a methodic and to-the-point account of the accelerated expansion that we witness in the cosmic world—from the inception of this phenomenon to its growing authenticity in the advanced field of cosmological physics.[3]

The briskness of space-time is forcefully announced in the operations of the cosmic world, where the galaxies approach the limit of haste. Beside the prodigious scurries in the folds of the universe, the standard rates of movements themselves can be bewildering to envision. The earth gyrates at 1,070 miles per hour (in reference to the equator) and hurtles at 67,062 miles per hour around the sun; the sun turns at 4,400 miles per hour and hustles at 486,000 miles per hour around the center of the galaxy; the solar system paces around the Milky Way with a rushing push of 568,000 miles per hour; stars in galaxy arms cycle about the Milky Way's center with the whopping rapidity of about 600,000 miles per hour.

Having marveled at the actuality of rapidness and acceleration in the plane of the cosmos, let us try to think through the possible role that dynamism fortified by acceleration plays in the reality of the continuum.

The Expediting Space-Time Pointing to the Underlying Tactics

The actuality of existence is gravely tied in with the deed of mobility; journeying is the only way by which the elements prevail. A perpetual mode of shifting is an ingrained feature of all the physical entities, including the living ones. Comfortably nestled inside our dwellings, we still remain a part of the grand dynamic push that goes on in the stellar setup. And if we were to leave the earth, we would have to gain an agitation independent of the earth's progression. In fact, the farther we go, the faster we need to fly (or brisker flying would occur), to sustain the intrinsic mobility in the nature of the reality.

The coexistence of the counteracting factualities—the unification by gravity and the process of breaking up via accelerated movement—can be grasped by envisioning a simple animation of a ball of ice shavings. Envision that this ball of ice shavings is made to spin with a forcible impetus, whereupon a constant and firm propulsion was given for the ball to continue to rotate (indefinitely). As this ball continues to whirl with rigor, the ice shavings strip away from the outer surface and, consequently, start to wheel vigorously around the center, counteractively balanced by the centripetal and centrifugal forces. Owing to the gradual dislodgment of the ice chips, the ball bit by bit loses its mass (although the total mass of the system still remains the same) (figure 11).

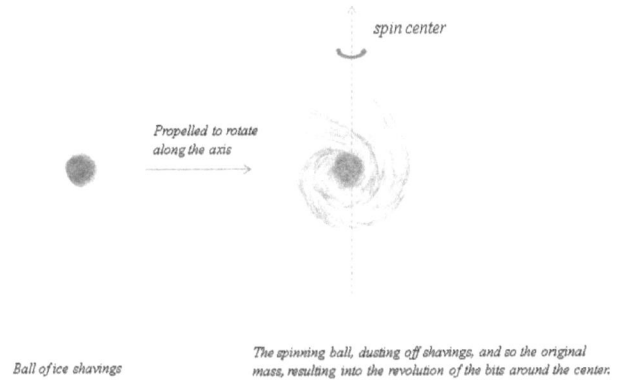

spin center

Propelled to rotate
along the axis

Ball of ice shavings

The spinning ball, dusting off shavings, and so the original
mass, resulting into the revolution of the bits around the center.

Figure 11. A spinning system, by the example of the whirling ball of ice shavings.

With the continuity of the spinning momentum, the ball eventually relinquishes every bit of the mass that continues to fly away via the rotary motion. However, the ejected ice chunks, as long as they exist, continue to be part of the basic center—the center of the rotating axis—simply because the entire system is enveloped by the elementary gravitational field and thus represents a network that emanates from a single body of the original matter. Therefore, the region encompassing the dislodged ice shavings, along with the center, exists as a grand unit. This example illustrates the existence of a grand unit through spins and rotations around the center—a grand unit that, although unified through gravity, constantly moves with increasing speed, where the distances between hurtling scraps gradually and rapidly grow.

The example isn't a literal depiction of what happens in cosmic evolution. As for the cosmos, it is rather a fully interactive field in which every crumb moves by the weave of universal forces. The example nonetheless brings forth certain subtleties of constant movement in a unitary thrust.

In the hypothetical event of a sudden cessation of movement, or of spinning, and thus of the rotation around the center, the uprooted ice shavings will drop back onto the original center—they have nowhere else to go. In this schematic, every bit of the dislodged ice shavings is itself a mass, although to a lesser degree, compared to the original ball. The envisaged simulation relates to the movements of planetary systems in a way that every shred of the cast-off mass—illustrated as dislodged ice shavings (figure 11)—will be possessed with its own spins, where the entire system acts in a singular makeup. Indeed, in a real arena a solar system is only part of the grander dynamics of galaxy, and thus its propulsion doesn't depend on only the internal core.

As we have noted, within the scientific acumen is buried the denotation that the moving cosmic bodies are held together by gravitational pull on one side and are withdrawn from each other as a result of the dwelling of yet another force, acting antagonistically to the force of gravity, on the other—the presence of dark energy has been argued retrospectively to be the force that is responsible for the ongoing inflation. As I noted above, these forces are realized at galactic scales, and the forces of dark matter—unseen matter considered to hold galactic structure—and dark energy are believed to exert themselves intragalactically and intergalactically, respectively. The bottom line, nonetheless, is that the cosmic world is enduringly expanding with successively aggrandized rates, and that the antagonistic force apparently is surpassing the forces of gravity—judgment based on its gravity-countering influence in the evolution of the universe and the flow of time.

In the felicitous design of space-time, the fruition of constant acceleration can be fathomed to be an indispensable feature; however, before that, we might be able first to appreciate how the behavior of rapidity might be rooted in the machination of gravity itself, that fallout under the notice of the observer.

The Case of Rapidity by the Spectrum of Gravity: The Necessity of Acceleration

The artificial space satellites stay put in a geostationary state about 22,236 miles above the earth's equator because their orbital periods exactly equate to the earth's rotational period. The satellite thus gets lodged as a fixed motionless object with respect to a defined point on earth. And although the satellite gains its own treading path, in a physical sense, it remains suspended alongside earth's movement; the earth and the satellite move as a single planetary body. The movement of the satellite is counterbalanced by earth's gravitational pull.

No sooner does a moving object leave the earth's gravitational pull—or, more precisely, no sooner does the gravitation between the earth and the satellite weaken—than the moving satellite falls into the gravitational clasp of another stellar object. The force of gravitation does not point to one-to-one traction between any two identities; rather, it attests to the occupancy of a field that permeates the dominion of materiality. For an entity, a decline of gravity in one sector is accompanied by its strengthening in the other. The setting up of the matter is concurrent to the imposition of gravity. This, in Einstein's order of relativity, implies that space-time is conformed through the gravitational force.

The nitty-gritty of space-time, in this respect, involves two opposing aspects. First is the presence of the gravitational field, and second is the installation of the interminable velocities that unvaryingly augment. As noted in scientific logic, the feature of exponential expansion that we are witnessing can occur only if a force antagonistic to the force of gravity also dwells emphatically in the capture of reality.

The combative association of gravity and antigravity, so to speak, can be seen as a consequence of the manifestation and continuation of actuality. Let me explain how.

In the shapely texture of binarity (or duality), gravity is welded to mass, and displacement to energy; both combined lay the fabric of the bustling universe. The permeation of gravity kicks off with the appearance of mass, the initiating texture of which can be described as an appearance of a jam-packed impenetrable singularity—similar to the dogma of "gravitational singularity" in scientific theorization, which advocates the existence of space-time as an infinitely dense singleton of mass prior to the occurrence of the big bang. Thus, in the sea of floating stellar systems, gravity is the reflection of that very force that interlaced the mass at the onset, albeit with a different intensity. Regardless of how large the universe grows, all-enveloped, it still acts as a systematic object that flutters in singularness, because again, it is the same force of gravity that holds the unsealed vista.

The movement, contrastively, inscribes the dwindling of the attractive drag, instilling the functional vitality in the spatiotemporal beat. Within the frame of the solo universe, the instance of singularity would denote the commencement, and reaching the limit of expansion would tab the termination. I may be able to substantiate this persuasion shortly, in viewing the phenomenon of beginning and end that occurs with respect to the observer. It is important to bear in mind, however, that in reference to the fluttering of the multiverse the cases of initiation as well as conclusion become inconsequential because the induction as well as the conclusion occur only in reference to the observer of that particular space-time history—thus the amplitude of the multiverse palpates endlessly.

In the grand scenery of mass and mobility, the recession of the two intergalactic frames can be seen as the gradual dwindling in the strength of the gravitational pull between them. And thus, in this supreme schematic, the progression of departure doesn't just follow a linear mode but rather accelerates. The lesser the gravitational pull, the farther the object travels and the higher the speed it acquires. The speeding in the cosmic arena can be reasoned to stem from the gradual decay in the strength of the gravity, and this doesn't necessarily imply that it is the dwelling of a conflicting force (the argued dark energy) that causes the gravitational abatement— we will look into this alternative possibility in just a bit.

Now, in the tapestry that breathes by the hammer of mass and movement, it is pivotal to apprehend that though the inception of the progression marks the beginning, for the flow to continue, the rate of progression itself must continually alter, enhanced. The lack of mobility and the fixed rate of mobility both, in reference to the flow, imply stagnation, a hypothetical situation that would oppose the streaming of space-time.

With a perpetual boost of velocities, the galactic space that recedes from us with exceeding impulses eventually would come to evaporate in wake of the dimming gravitation. It may stir a paranormal feeling, but the fact is that, in the absence of the gravity, they won't even exist. It is not that they fly far off to some unknown place but rather that, once the velocity of an object surpasses the strength of gravity, the object ceases to be. If you see it, it is there, and if you don't, it is not. Let us muse over this causality from the vantage point of the observer.

The Case of Rapidity in the Spectrum of Gravity by the Coupling of the Observer

The starting point implies the initiation of the movement against the pull of gravity—it happens to every mass-energy form that exists anywhere in the field of the universe. Now, the hikes in the velocities within the universe that is closed and self-contained can be seen as the widening of space between the entities, analogous to ballooning, where the only way the entities on the balloon's surface can progressively move away from each other is if the balloon itself is constantly blown bigger (this is a part of the current scientific view to accommodate the fact of expansion). By accounting for the presence of the watcher, however, on the other side of space-time we may be able to assimilate not just how (and why) the expansion occurs in the universe that is closed and justly tied but also why the object ceases to exist and does not wander off. Although both situations, the expansion of the universe and the petering out of the object, are equally possible in the wake of diminishing gravity, it is the plot of space-time abatement, rather than inflation, that seemingly fits the scientific paradigm in offering an ultimate schematic encompassing our own placement.

Let us now try to ruminate over the foundational meaning behind the expedited opening out in the vista of the universe, with the perspective of a watcher that remains behind the scenes. As follows, there is yet another way of picturing the underlying synchronicity that goes on between the forces of gravity and the force that is responsible for the commencement of the dynamic bearing. The dwelling of the watcher, the observer that beholds the functionalities of space-time, must be envisaged for its joining-in in successively magnifying the force of departure. With real identity being the observer, the perpetual acceleration of the dynamism can be seen to correspond to the dissolution of the elements, space-time—the gradual fading off of the matter. The idea may seem a little outlandish at this point; we will have more on this in a while.

In the grandiose dreamland of space-time from the perspective of the watcher, the movement, as we have acknowledged, would be nothing but the serial arrangement of discrete snapshots, and in the flow of the time, the discrete snapshots appear momentarily, pinning a time point, to disappear conclusively. Every time point is unique and is established by the compounded interactive field (as noted in chapter 5, "The Essentiality of Annihilation in the Pulsation of the Reality"). Thus what appears to be a colossal expansion from within the superstructure, from the view of the watcher would be seen as the gradual dissolution of the materialistic plane. And in this framework of observer and observed, the ongoing executions entail only coming into view and then disappearing—whether refereeing the moment-to-moment progression or considering the stream of the worldsheet in totality—to fade out altogether, mirroring the diminished gravity. Let us try to rationalize this on the grounds of scientific notation.

The watcher that correlates with the mathematical empty space—for it in itself is a propertyless state and remains beyond the possibility of measurement—is the observer of the functionalities. Thus the abiding extension in the expanse of the universe would reason the equivalent stretching in the arena of the observer too. How can the expansion in the lodging of the observer be reasoned when the observer nests outside the limits of the measurements? Equally important is that from scientific erudition as well we know that there isn't any vacant space outside the ambits of space-time; the universe doesn't expand into a preexisting space, and the circuit of reality stays put, closed and compact. Hence, the pedantic findings along with the reasoning from the perspective of the dimensionless observer suggest that, instead of the universe itself going through the adventure of speedy expansion, it is the materialistic arena that first manifests, then etiolates, to bleach out all the way, in reference to the observer.

The ongoing transformation that is an essential component of the vitality occurs only by the succession of manifestations and the fading out, in the scene of the observer. The more the observer is realized, the more the speed (the rapidity of fading) is gathered. This point can be comprehended better in context with the topic of dreamy and awakened states that we will come upon in the last chapter. From within the space-time framework of relativity, though, the phenomenon of "fading out" will be perceived as a hastening in departures. For the space-time that is closed and compact, the scientific outlooks too can be seen to advocate the continual and gradual dwindling in the force of gravitation, which consequently would lead to the complete evaporation of the matter. This isn't in the least mystical, because the fading out would occur only for an observer. We will come to this again.

The American astronomer Edwin Hubble (1889–1953), known for his efforts in initiating deep-field astronomical exploration, systematically pointed out that the recessional velocity of the galaxy increases with the increase in the distance relative to the earth. He indicated (in 1936) that all galaxies, in relation to the Milky Way, have positive redshift—a way to deduce that a stellar object is ebbing. This judgment is based on the fact that the light that reaches us from a far-distant receding cosmic object exhibits a proportional increase in wavelength—becoming redder or shifting toward the red side of the spectrum. The astronomical redshift, however, is rather seen as a consequence of stretching—of the waves—owing to the expansion of the spatial extent itself, directly rendering the increase in the wavelength, or the redshift. However, the phenomenon of the redshift will occur equally effectively if the galactic body gradually fades; it will proportionally radiate at a lower frequency and would stand to be redshifted. A receding celestial cluster will show the redshift as much as a paling or disappearing galactic bundle would. And under the umbrella of the observer, the steady vanishing of the elements as well as the opening out in the spread of space-time would emerge as one and the same phenomenon. However, from within the space-time congregation, the instance of ebbing would be enacted as an episode of flying off (from the range of gravity).

Any materialistic cycle, whether we talk of the ephemeral vibration of a quantum particle, the life story of a conscious embodiment, or the oscillation of the entire universe, in its sweep, incontrovertibly connects to an initiation on one side and termination on the other. However, neither the elements nor the laws start to flow unless they are accompanied by a capacity that notices them. Thus, in this framework, within the myriad floating universes, the inception of a particular space-time coincides with the witnesser witnessing it. For this compact and closed universality, its arrival (the mass) and progression (the movement) are noticed on equal footing, emblematizing the commencement. And from the stage of the observer, the culmination of the progression and so the disappearance of the elements (the mass) would bring up the termination of that particular physicality.

In other words, the inception of the vitality—that is, the domains of conscious reality—can be envisioned as the beginning of the thawing of duality between the physical plane and the subliminal fold, for there dwell two distinctive windows in the plane of the observer (as we have noted in the preceding chapter, under "Peephole on the Behaviors of Wave and Particle"). And as we have acknowledged, the distinct windows—the neural and sweeping view of the observer—both are parts of the one and the same tapestry. In the scene of the observer the two reflections of the same mold are endorsed.

The Mode of Self

In the hierarchy of multitudinous living organisms, it is the human that apparently beholds the intellective and supraliminal attributes with greatest perception. A few years ago, I ran across an appealing talk titled "Mirror-Mirror," given by Mark Pendergrast. The topic, in a nutshell, exposed the credibility of mirror as a utility in reflecting the attribute of being self-conscious. Informatively, I learned from this talk that in the course of animate evolution, the characteristic of being-conscious-of-the-self starts to emerge in the species of apes, for apes can relate to their own reflections in water.

The presence of the powerful sense of self in the human, in retrospect, highlights that in the sentience of the human the sense of identity has been shifted, at least partially, from the feeling of being a set of elements to the awareness of being manifested as the watcher. The more the self is witnessed, the more the autonomous recognition of the watcher follows. In the dreamy landscape of the universal forces, we thus remain fractionally diagnosed as the watcher, albeit unknowingly. And in that sense, in the evolutionary ladder of the animate expansion, the ultimate stage would be the full realization of the truest identity, which only witnesses the elements of the universe, within its bounds.

By thermodynamic commentary, the act of *living* announces the calibrated balance between the bounds of enthalpy and the freedom of entropy. The two polarities of this counterbalanced state are as follows: (1) All-enthalpy, a completely rigid (bonded) state, meaning that in the field of reality the watcher is least recognized. Obviously, that cannot be a living entity, for the observer is an essential component in the streaming of space-time. (2) All-entropy, meaning that all bonds are unleashed, thus the total lack of influence by forces of the universe. This cannot be a living state either, for it is the ingredients of space-time that constitute elements to be observed. It is the blend of both—at every given time point—that would make conscious life. And it is the consecutive transformations between enthalpies and entropies that would propel the time and the vitality.

CHAPTER 12

$\exists x P(x)$

Peppered Space-Time:
The Dabs of Units

The Tearing of the Continuum

Envisioning the Units of the Unified Field

The appreciation of how the space-time unit is constituted is elemental not just toward seeing the complete panorama but also in rationalizing certain empirical quirks. A unit of the vital universe isn't just the smallest embodiment of the obvious reality; it also makes room for the dynamism that plays in the ticking of the biological division. For one thing, to say the least, it is the assembly of the biology only that stumbles upon the display of the universality, to begin with.

In the systematic network of living embodiments, the hierarchical array of intellections renders increasing levels of perceptivities, and on the evolutionary spectrum, the vertebrate division carries the sense of the intellective attributes to the greatest extents. It is by this very packing array along the way that runs another deeply interlocked orderliness—from the state of being conscious to the state of becoming-the-consciousness, with the concomitant timeless realization of the watcher swept beyond all. It is in accounting the fact of consciousness into the framework of reality that we can realize the dabs of real units that flutter like exclusive occurrences in the endlessness of hyperspace.

The fluidic unit of reality is a cocircuitry of cognitive and tactile domains, cooperatively effectuating the dots of time, which in totality surge as a cast of a life-span. The surfacing of a life-span in the arena of reality brings forth that unit that, although it represents universality on its own, dwells among the congregation of the numerous.

The unified field encompasses the parcels of the subliminal too. Once all the sections of the assembly are fittingly tailored into a whole, the unveiling of the ultimate megacosm ensues, by the texture of that very formulation. The inclusion of the mental attributes in the scheme of reality has been a matter of oblivion, conceivably amounting to the twist that the physical reality in itself turns out to be a phenomenon of the inner weave, where we peer into a field of physicality that per se stays curved through the poses that the scope within only vibrates. The averments of quantum and classical physics fastidiously depict the terrain which, after we have accounted the subliminal carriages and our own positioning in the universe, reduces to a scene of warped cascades for every given observer. This does not mean that the reality being a case of internal panorama makes the collective any less significant. The sweeping province of peculiarities still would prevail as the cogent reality, for it is in this terrain only where the laws of the universe stubbornly carry on; it is where the space-time continuity prevails; it is where the acts of embodiments and the cessations through annihilation occur; it is where the desires come to pass and the intuitions are seized.

Working Out the Quirky Facts: The Authenticity of Rips and Tears in the Fabric of Space-Time

The avant-garde reasoning in the formulation of the way the quantum world behaves paints that space-time girdles the instances of tears and ruptures in its own flow of continuum.[1] The advanced outlooks, involving works of Shing-Tung Yau, at Harvard University; Gang Tian, at MIT; Brian Greene, at Columbia University; Andy Lutken, at the University of Oslo; and Paul Aspinwall, M. Ronen Plesser, and David Morrison, at Duke University, show that the mystifying transitions involving severing and puncturing do befall in the bed of space-time (figure 12.1).[2] These incidences of disarray at first sight may sound catastrophic, but in the boundless fields of matrices, they can be seen to be the episodes of a cardinal value—imperative for the sustainment of the athletics that go on in the absolute sphere. Before appraising the functionality of the rips and tears that prevail boomingly, we first need to figure out how they come to transpire in the first place—how the break in the continuum comes to pass. Putting the task the other way around, we can strive to define a field in which the continuum, instead of being broken and then repaired, is sustained by the streams of units, where the appearance of gaps and fissures is an automatic outcome.

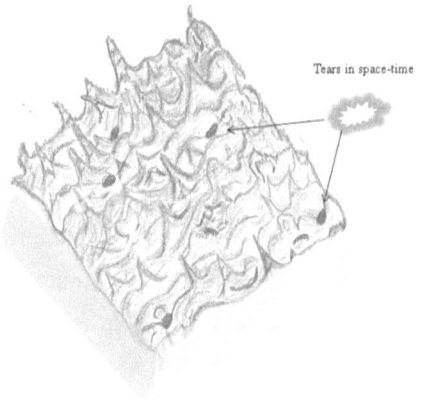

Tears in space-time

Figure 12.1. The instance of fissures in the quantum depiction of space-time

In the mosaic of actuality that takes up billows, furrows, stretched bumps and drops (from quantum interpretations), the appearance of lacunae has been mulled to be a common occurrence, discerned through the medium of mathematical abstractions in constructing the unified theory. The interruptive nature of space-time that incorporates the instances of breaks and holes has been predicted by the mathematical tapestry of the string theory—technically addressed as flop transition, the instance of tearing followed by having the cut glued back together again, was shown to be tidily hemmed in this model.

With the jelling of the continuum in the form of units a possibility, the fissures in the jungle of these units doesn't just become a causative natural occurrence; their permeation comes to be a linchpin in tying down those loosened units to the feats of the "beginning" and the stunts of the "end."

Defining the Unit of Actuality: Taking Indubitable Clues (Again) from the Guise of Mathematics

It is again in the models of mathematics that we can see the painting of space-time in units and consequently understand the significance of discontinuity in the spread of the Gordian continuum. Alongside wide-ranging applications in social sciences, chemistry, and pure to applied mathematics, the mathematical asset of the polynomial function sleekly serves as a prototype in envisaging the unitary expressions of the space-time continuum. The polynomial is a mathematical statement of variables, coefficients, and constants, phrased using addition, subtraction, multiplication, and a nonnegative integer exponent. The variable in the polynomial must not carry a negative exponent, and it cannot be a divisor. A few examples of polynomial statements are as follows:

$$4x + 8 \qquad\qquad \text{(i)}$$
$$x^2 - 8x + 7 \qquad\qquad \text{(ii)}$$
$$x^3 + 5x^2 + 6x \qquad\qquad \text{(iii)}$$

The level of the exponent is called to be the degree of that expression. The first expression does not contain any exponent—it is a degree 1 expression. The second statement has the exponent of 2; thus it is a degree 2 polynomial, also termed a quadratic function.

The third expression carries the power of 3; thus it is a degree 3 polynomial, also called a cubic expression. The graphs of these mathematical verbalizations are unique outcomes, stemming from the level of the degree the equation holds. Graphing these expressions would mean depicting their numerical mesh in a two-dimensional x, y complex plane—a Cartesian plane that has $-x, +x, -y, +y$ sides (figure 12.2).

The expression with degree of 1 gives a linear graph, typical of the polynomials that do not pack an exponent. The graphs of the polynomials of degrees 2 and higher are always smooth, continuous, nonlinear curves—a quadratic function will always be a parabola; a cubic function will always display two flexions, twice switches in the direction of movement; a quartic (or biquadratic) function always develops into an undulation that has three deflections or bears three switches in the mode of growth (figures 12.2; 12.3). The degree defines the graph's behavior, and the smooth continuum of the graph designates it as a polynomial; signifying that in this averment, the constituent elements are there *only* to be interfused into a composite, which in totality represents a closed autonomous entity. We will have more on this shortly for a clearer picture.

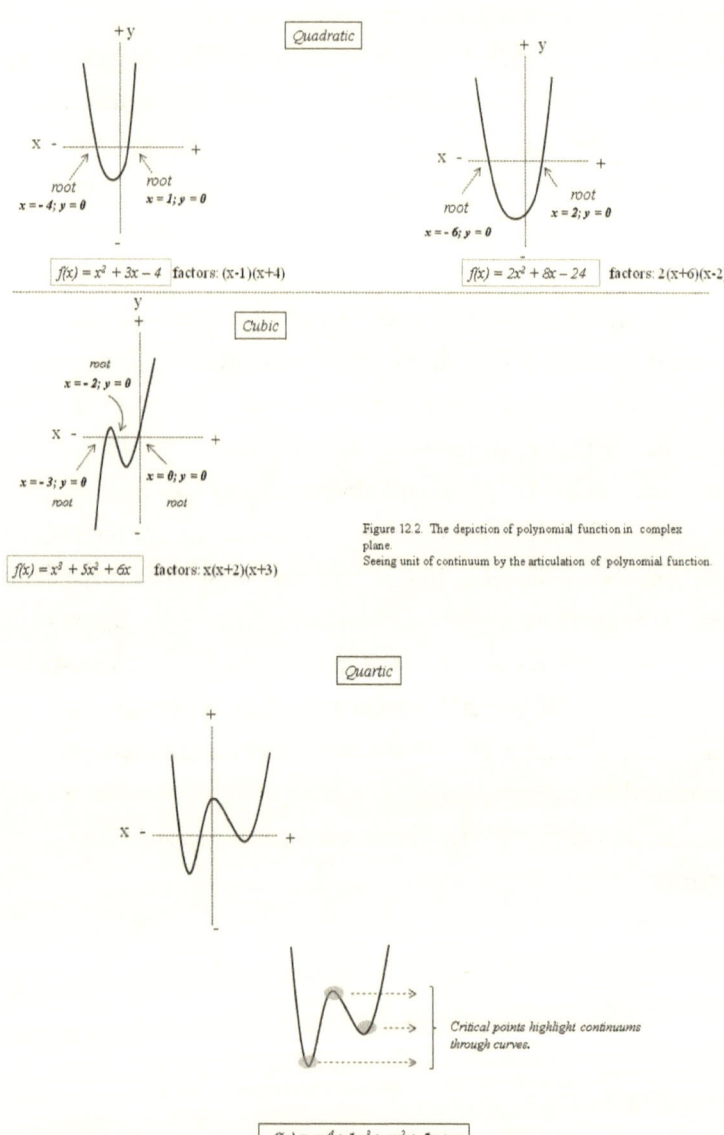

$f(x) = x^2 + 3x - 4$ factors: (x-1)(x+4)

$f(x) = 2x^2 + 8x - 24$ factors: 2(x+6)(x-2)

$f(x) = x^3 + 5x^2 + 6x$ factors: x(x+2)(x+3)

Figure 12.2. The depiction of polynomial function in complex plane.
Seeing unit of continuum by the articulation of polynomial function.

$f(x) = ax^4 + bx^3 + cx^2 + dx + e$

Critical points highlight continuums through curves.

Figure 12.3. The articulation of a polynomial could reveal how the unit of continuum is procured. A quartic polynomial is constituted by welding four subunits in the continuum. These subunits survive only as a part of the continuum, while the entire function fruits only through interfusion of the subunits.

Grasping the Unit of Space-Time by the Voice of the Polynomial Expression

Sketching polynomials in a complex x, y plane (i.e., establishing the correspondence between x and y) requires the polynomial to be first decomposed into constituent bits (a procedure known as factorization). The bits, when multiplied, give back the original expression. The polynomial (or the multinomial) form implies an autonomous numerical phrase, and this means that the phrase as a whole equates to zero. Thus, the preceding expression (ii), $x^2 - 8x + 7$, in equation format would be $x^2 - 8x + 7 = 0$. And because the equation as a whole equates to zero, each constituent that we get by the act of factorization also comes to be equal to zero. So, for a multinomial equation $x^2 + 3x - 4 = 0$, the factors $(x - 1)$ and $(x + 4)$ both would match to zero $(x - 1 = 0; x + 4 = 0)$. In the complex x, y plane, thus, 0 on the y-axis, for this rendition, would correspond to 1 and -4 on the x-axis. The act of factorization paves a way to see the complex behavioral traits.

The values on the x-axis that correspond to the 0 on the y-axis (i.e., 1 and -4 in the preceding example) are known as roots of the expression. The prior example carries two roots, and in the complex x, y plane, the roots show us that the x-axis positions 1 and -4 tally with the y-axis point 0, in charting out the complex behavior (figure 12.2). The number of times the graph hits 0 on the y-axis is rooted in the level of degree that the expression packs. If a graph encounters zero on the y-axis three times, it is a graph of a cubic expression (figure 12.2). Similarly, if the graphical sketch crosses 0 on the y-axis four times, it is the articulation of a quartic nature, bearing the exponent of 4, and so on (figure 12.3). The U-turn that eventuates in the behavior of the graph is mathematically termed as the critical point. Figures 12.2 and 12.3 illustrate that the quadratic has a single critical point; the cubic expression has two, while the quartic (or biquadratic) has three. The number of critical points, hence, is always one less than the degree of that intonation.

Polynomial expressions have a wide range of scientific applications, such as in modeling growth curves, in mechanical estimations that involve slopes and gradients, in atmospheric sciences and meteorology, and in estimating economic and business trends, to name a few from their overarching presence in the vast bounds of chemical and physical sciences. Our purpose of discussing polynomials here is mainly to catch sight of the continuance that exists discretely in the ultimate field, where the ruptures and tears are not only allowed but are systematically enforced in the engendering of the utmost vista. The cluster of continua persists through the mosaic of discontinuities.

The articulation of the multinomial avers the cast of an autonomously functional unit that stems from the coordinative relationship between the constituting elements. The terms (e.g., $4x$ and 8 in expression (i)) of the polynomial smoothly sew together to beget a unit of a viable continuum. And although this unit describes an order of perpetuity, which per se emerges interactively, from beyond its own bounds it emerges as a discrete piece. And despite that all of the constituent terms justly participate in the declaration of the deportment, their holdings in aloofness do not amount to anything at all.

In the game of this mathematical texture, the polynomials are executed only when the variable, x in the earlier examples, does not stand to be a divisor or carry a negative exponent. For example, the statements $4x^2 + 3x - 8$ and $4x^2 + 3/x - 8$ do not prescribe the tones of polynomial articulation, for the reason that the first declaration carries a negative exponent and the second one bears a variable as a divisor. Both the potentialities—the occupancy of a negative exponent (on the variable) and the variable being a divisor—lead to the failing of the process of unbroken interblending, which is essential in setting up the functional continuance. A polynomial articulation emblematizes a functional peculiarity and, in the prior example, illustrates the very potentiality of the x itself. We cannot use the variable x as a divisor, because that will cause a disjunction in the flow of the function.

For instance, for an expression $4x^2 + 3/x - 8$ charted in the complex plane, the value for the y-axis that corresponds to 0 on the x-axis cannot be estimated; nothing can be divided by 0 (see chapter 9, "Missing Undulations in the Cosmic Regimen"). Elements cannot subsist in a situation where the permeation of zero is discounted, and here it implies that the function of x cannot survive when the x itself becomes the cutter and, thus, introduces a halt. No sooner does the instance of stoppage occur than the disappearance of the expression altogether takes place. Likewise, the variable with a negative exponent would effectuate the fractional realm, and so again, the execution causes the encounter of discontinuity in the flow of the whole; the by-products of negative exponentiation image the wholes but do not meet them (see chapter 10, "The Duality of Bidirectionality in the Hammer of Reality"), again causing the total dissolution of the function.

In the continuance of the function, the factors sort of represent the subunits that do not stand on their own—the way subunits pose in the layout of the power set. The framing of the intercommunications in the fulfillment of the multinomial simulates the underlying mechanism by which the flapping units of the physical continuum come.

A little more detailed look into the machination will aid in appreciating the shades of reality that the polynomial can offer. A quadratic expression:

$$2x^2 + 8x - 24$$

The first step in the factorization of the preceding quadratic expression is to ascertain a greatest common factor (GCF) in order to do away with that which is dispensable, for determining factors has to do with finding out the interrelationship reflected on the complex x, y plane. After taking out the GCF, the above equation becomes the following:

$$2(x^2 + 4x - 12)$$

The factors of the expression show that the roots for this function are -6 and 2, from the following:

$$2(x + 6)\,(x - 2)$$

Solving $2 * (x + 6) * (x - 2)$ (i.e., multiplying 2, $(x + 6)$ and $(x - 2)$) gives back the original statement $(2x^2 + 8x - 24)$. And because the multinomial diction equates to zero $(2x^2 + 8x - 24 = 0)$, the constituent factors will be held equal to zero as well. Hence the factors from the preceding mathematical statement are recorded as follows:

$$(x + 6) = 0$$

$$(x - 2) = 0$$

As we grasp from the previous example, here the two factors $(x + 6)$ and $(x - 2)$, since both equate to zero, highlight that in a complex x, y plane, the −6 and 2 positions on the x-axis correspond to the 0 locus on the y-axis. Sewn together, the factors cook up the peculiar function (figure 12.2).

The visual representation of a cubic declaration that engages factors x, $(x + 2)$, and $(x + 3)$ is shown in figure 12.2. These factors declare the rise of three elemental subunits in the window of the complex plane. The two critical points of the depiction epitomize a smooth, frictionless association of the subunits that are interfused with each other (figure 12.2). Similarly, the three critical points in a quartic diagram bring attention to the unwrinkled sailing of its four subunits, which are sewn in continuum (figure 12.3). A quartic expression will have four roots, and therefore, in the complex plane, the function will stem from the four subunits (figure 12.3).

These functional assertions are allegorical of the units of the continuum, not just because they stand alone but also because the distinctive attributes here—signaled by the factors, or the terms—interblend to synthesize a steadfast continuum. The equitable integration of all the subunits generates an autonomous entity that, although it engages the elements in the continuum, in its entirety hints at a dwelling of a discrete unit. A functional unit is a stand-alone piece made out of dissimilar subunits that reside in boundlessness.

This kind of interactive unit in the nature of reality stems from the weld of all that we run into, both at the bodily and the clairvoyant levels, on the stage of life. The snug admix of two together in a nutshell brews a unit of the continuum. The continuum of subsistence, borne by an individuality, translates into an entirety of continuum—functional sequence that is tied to a beginning on one side and to an end on the other.

The Peppered Space-Time Continuing by the Anchor of the Observer: The Subsumption of Rips and Fissures

From the vantage point of the watcher, the elements of the all-inclusive space string over and above the boundary of the perceived external and cognized internal. And in the formulation of the continuum, the coming about of the directionality, as well as the duality, concurs with the permeation of the five senses that cause the projection of an askew plot in which the quasi case of extrinsic and intrinsic crops. And in a field where only the neural perceptions are seized—that is, apart from the field of the watcher—the scene of relativities that manifests through us, and between us, goes eclipsed. Tightly hemmed in the cogent scheme of the physical reality, and being validated time and again, it is the certitude of the same erected relativity that would be seen to be pitched thoroughly from the view of the watcher, beyond the neural sensing.

In attempting to concoct the schematic of the continuum, or contrive the nature of the ultimate texture, the factor "us" is usually overlooked. The citation of us, here, speaks of the elements that permeate the visceral levels—the materiality of the mental plane. In the opulent dynamics of the universal forces, both the facets of the body–mind continuum would be held on equal footing, for every twinkle of time, akin to the continuum of space-time. The functionalities that come to pass in the biological arena are the part of the ultimate tapestry much the same way the intuitions and cognitions are—all borne by the interchangeable tide of the continuum. The truest identification, the watcher, remains in oblivion owing to back-to-back lining up of the subunits that go on sprouting in the perturbative landscape, the way string theory announces (chapter 6, under "Space-Time Elasticity and the Integrated Picture from the String Theory"). And even though all the activities are occurring in reference to the watcher, and it is the watcher that is advancing, unfolding, or even going through the life and mortality, yet it is again the watcher only that remains unaware of the knowledge of its own presence.

Thus, in the paramount fabric of make-believe, the elements of the universe are seen as the elements of the self, especially the ones that reside in the subliminal. And no sooner does the image of the self echo than the manifestation of others pops up. In this dreamy field, it is this fabricated sense of identity by which the streams of parallel universes are effectuated too, for every given time fraction. The pounding of the parallel universes is an unconditionally genuine phenomenon that quantum mechanics declares unfailingly, but the case of parallelism can be seen to throb only after subsuming the conscious manifestation, which on the other end takes the pitch of the observer. From the vantage point of the truest over-and-above-self identity, the watcher, the parallel universes would be seen as the tides of discrete continua, shifting and altering for every beat of time, as we learn from quantum mechanical views.

As we have acknowledged previously, the beating of the parallel universes has to do with the watcher witnessing some of the elements out of the large solitary universal set of elements (chapter 9, "Singling Out the Core of Nonduality from the Grand Frame of Duality"). As the time continues to flow, the new elements would emerge interactively. So although in the flow of the parallelism any given prevailing vitality witnesses only a fraction of the universal set and remains self-confined, the flow of time occurs endlessly, constructed by the units of the continuum.

And although there resides only one universal set of

functionalities, and the forces that set up their stage, the assemblage of the parallel universes does not echo the instance of a unit, or an entirety, owing to the lack in the continuity of space-time that the parallel congregation features. The multiverse of parallelism takes up the ruptures and tearing in the cast of space-time, which comes in sections.

Circuitries in the wheel of life remain molded in concatenation, analogues to what we see in the lucid example of polynomial expression. A sole animated articulation thus imparts an act of continuum in the higher field of compounded circuitries. In this collective field of surpassing dimensions, not only is the continuum sustained, but the phases of initiations as well as of conclusions are met.

In the field of the hyperspace, the widespread culmination of the holes, tears, and ruptures, thus, is not just some nugatory outcome; the disruptions that these punctures highlight are, in all respects, constitutional to the bubbling of the multiverse and the functionalities through which we all subsist. The hyper-field, where not just one continuum but the multitudinous continua pulsate, purports a perturbative field that not only strategically accommodates all elemental bearings but also allows the manifestation of the two ends of the cycle, the initiation and the termination, in the flow that by and of itself abysmally continues.

CHAPTER 13

The Conception of God by the Twist of the Human

*Paraphrasing the Mass-Energy
Conservancy across the Breadth of Time*

The Infinite Voyaging but No Escape: Conserved Mass

Unveiled by French scientist Antoine Lavoisier, in the eighteenth century, the principle of mass conservation, primarily established by the mode of chemical reactions, suggests that for a closed system, the amount of mass abides conserved over the passage of time; it is neither created nor destroyed but just undergoes rearrangements, or transforms between divergent forms. For a chemical reaction in a test tube, the mass of the reactants breaks even with the mass of the products.

Weaving the picture of the conservation in the advanced principles of special and general relativity engage intricate decipherments, mostly for the reason that the relativity theories account space and time as a single body of actuality, and consequently, the flow of time need be subsumed in the manifestation of matter, and matter cannot be subscribed to solely. In the doctrine of Einstein's special relativity, the equivalence of mass and energy ($E = mc^2$) tells us that mass is energy, and vice versa—all types of masses blazon forms of energy, and all descriptions of energies echo stamps of masses, persuasively uttering that there occurs two types of mass: rest mass and relativistic mass. The same multiplicity is held by the nature of energy too: the state that equates to mobility and the one that corresponds to inactivity. By classifying the energetic attributes in correspondence to the framework of mass, for an isolated and closed system, the principle of mass conservation is seen to hold for a given observer—that is, in the structure of Einstein's special relativity.

The envisioning of mass conservation by the theory of general relativity—that is, subsuming the gravitational component—is, however, a complex conundrum. Intricate mathematical subtleties

are involved in seeing the perplexity of picturing the mass conservation in the formulation of general relativity.[1] For our purposes, nonetheless, the implementation of mass conservation in the
code of general relativity is an order that brings up an illuminating
essence.

Welding mass conservation into the landscape of general relativity can be achieved only when a point of reference of an observer
is accounted for in the plot. The theory of general relativity blends
the gravitational field into the dynamic arena of mass and energy,
and the riddle strikes as soon as we want to weave gravity into the
perpetually expanding universe, which must occur in reference to
an observer for us to value the conservation—the notation may
appear a little disconcerting, but we will get back to this in just a
bit. Here suffice it to say that a feature of mass-energy conservation
as an average system property cannot be deemed possible, in actuality, and by the voice of general relativity. Seized by the scientific
findings, however, the percept of the conservation of mass (and
so the energy) for a holistic system is not only intuitive but also
appears to be the only way.

In the photoplay of time, the logic of preservation thus advocates that the mass as a bulk has always hovered in the exact same
amounts and will forever be that way. Because mass is energy as
well, the amount of energy subsists on equivalencies too—only
to be transformed from one form into another, never getting compounded or reaching a culmination. *Ex nihilo nihil fit* is a Latin
verbalization for "nothing comes from nothing," which was first
reasoned by Parmenides, a Greek philosopher, in the fifth century
B.C., in rationalizing the dogma of secured conservation in the order of the cosmos. It was associatively believed that nothing could
be either created anew or destroyed altogether into nothing—the

affirmations that currently tally with the scientific hold of the conservation of mass and energy.

Before going deeper into the purity and beauty of the quantitative conservation over the breadth of time, let us first pin down what the quantitative flow actually implies for a system that remains closed and compact and, as noted earlier, where space and time stay interfused. Although the progression of time is captured as new articles coming into sight and older ones leaving, in the ambience of an enveloped physicality, the succession of time translates into the routines of transformation, without shedding or putting on any weight. The jets of flow and ebb, or the literal *movement*, cannot make the scene within the furls of an encasing that is self-invoked and thus warped, where space and time manifest only by occupancy. There exists no interelemental gap of any size. Routine of transformation is the only way forward for a system that is autonomous and closed.

A few words on the scientific background of spatial continuity in the universal framework would help here (and we have seen this earlier in Chapter 2, under "What is Mathematical Gap?"). The insight that the stretch of the spatial continuum stays put free of empty space emerged from the interpretation of Heisenberg's uncertainty principle, suggesting that the spatiotemporal plane is exhaustively permeated by the makeup of matter, or its uncertainty; that is, each element possesses equal probability of being discovered anywhere in the whole infiniteness of the macrocosm. And from the current models of physics, we have learned that space-time exists as a field, where the matter and forces coordinate harmoniously, stating the case of exhaustive occupancy.

An obvious skepticism flares up. If the self-sprouted spatial sheet is packed and sealed all the way, how do we perceive behemoth intermatter's unoccupied extents in the stage of the cosmos? The explanation may again lie in the way our neural faculty seizes the orchestration of relativity. With regard to the interobject voids that meet the eye in the plane of the universe, the appearance of the emptied stretches of the interobject sweeps may refer to a quasi-picture that pops up as a result of the established materialistic relativity in reference to our neural sense.

It is from the perspective of the matterless observer that the space-time stretch would literally manifest, without the involvement of any interelemental gap. The expanse thus hovers compact, where the attributes of matter stay jam-packed into a systematic bearing; the absence of interspatial vacancies does not imply that the spatial sheet is all the way stiff and tight, for it is the equity of the matter and the fundamental forces that induce the ultimate texture, where the ply of space-time unceasingly remains aqueous through all its confines.

The watcher that accompanies space-time is a measureless attribute; thus the expanse of the watcher's (observer's) extension is decided by the measure of that which metamorphoses and maneuvers—that is, the materialistic elements.

With this prologue, we may now be able to picture the mass conservation all the way, which can be seen to be effective against the reference point of an observer.

The Mode of Conservation in the Infinite Ambits of Parallel Worlds

The principle of conservation that fits well in the scheme of parallel worlds holds to be accurate only for a given dab of space-time—that is, for a unit-of-continuum (the way we noted in the previous chapter) that remains encased in the field of an observer. And in the flapping of the absolute cadency, where the multitudinous such cocooned patches effectuate the shadows of the parallel universes (the way we acknowledged earlier, in Chapter 9, under "Singling Out the Core of Nonduality from the Grand Frame of Duality")— although for a given observer the net totality of matter remains the same—in between the scaffoldings of the parallelism, the bulk of mass-energy would not just appear to carry different volumes; it can also flash ambiguous shifts in proportions. Thus, in a nutshell, it is not the case that the amount of mass-energy does not shift. Ultimately it would. For one thing, the continuum of physicality occurs through the prevalence of the parallelism, which does not purport the implant of a sealed and warped entity. And it is by parallelism that the continuity of the time is maintained as well. The gesture of the bona fide time does not put on the marks of beginning or end; the time, come what may, cannot start or finish but can only cruise. The geography of inception as well as of completion becomes visible only under the arch of an observer, for a given measure of continuum in the field of parallel continua.

The closed and isolated conformity in the ultimate texture is a unit-of-continuum that makes the scene by the blending of the palpable, the psychic, and the observer, which notices the succession of the phases in the course of the world line. In the whipping up of the corporeality, the stances that are of outright prerequisites are (1) the tokens of the beginning and the end and (2) that there must dwell "something" rather than the oblivion of nothing. The coming about of both of these conducts can be reconciled into the framework of the ultimate beat only when these episodes are seen to be occurring in reference to the observer—the observer that is alive for, but positioned separate from the influences of, space-time. The outset of time or the surfacing of space cannot be nailed down unless they are pictured within the warped scope of the unaided bystander.

By the same token, as we have appreciated earlier, the full awareness of the emotions and the implant of an intuition cannot be deemed plausible without the reference point of an unregimented witnesser being taken into the rationalization. The constancy of a living thing through the successive alterations of growth and transformation, and by the cycles of sleep and wakefulness, again cannot be explained without accounting the encasing presence of an unfettered watcher. And, over and over, it is only by the blanketed omnipresence of the watcher that we can envision the streaming of nature that goes on endlessly—effectuating a scene of eternity, the cast of no-beginning and no-end, and that something always palpates, by the hyperscopic ambits of the multiworlds.

The Act of Progression in the Tightness of Crammed Space

As briefly noted above, in the extent where the congregated matter lie snuggled, disallowing any intermatter room, and so any hiatus or breach, the task of dynamism is fulfilled by the train of transformations; there is simply no room vacant to efflux or crane. Thus, in the befitting texture of the materialistic arena, while the expansion denotes waning, the "journeying" stands for transformation. And so, in the full-blown scheme, the universalities are endlessly subjected to the pull of transformation, as well as to the mounting and waning, in spewing the specks of time. However, to our neurological—or telescopic, for that matter—windows, the succession of transformations would transpire as the passing of time; and the stunt of waning, as the inflation of space.

In the architecture where the array of mass-energy forms stay snuggled and closed under the ambience of witnessing, neither the locomotion nor the drawing out can be regarded plausible. The ongoing transformations and the matter diminishing into a state of insignificancy are the two modes by which unabated effusion of space-time can be conceived. As we have appreciated previously, from the vantage point of the watcher, the series of transformations would crop up as shifts in the population, and the stance of waning to the point of vanishing altogether reveals as the gradual unfolding of duality. The catch of duality—that is, the discerned acknowledgment of inner from outer, space from time, or the bodily from the subliminal as two aspects of the same continuum—would come to sight only in the vista of the watcher: the lucidity that, although pointing to the equivalence of mass and energy, does not occur in the apparent actuality of the physical setup or to our neural senses. In concealment, the transfiguration is the only mechanism by which space-time reverberates, where the ebbing and annulling is an essential aspect in the execution of the ongoing transformations, with unending cycles of breaking down and re-forming; there is only so much material in hand.

The full-length trip of a creature in the fabric of the universe endorses the authenticity of mass-energy conservancy.

The Echo of Creature

Nothing stays permanently except for sureness of the trait—watching, that is, with respect to a sentience of reality. In the continuum of physique and intellect, the external and inner planes would constantly reflect under the permanence of the observer (figure 13.1). Thus, the presence of the watcher on the other side of space-time isn't just interfused with the debut of the matter; the witnessing is intermingled with the instances of dematerialization and re-formation as well. This is not to say that the presence of the watcher causes the concoction of physicality or time; the mechanism by which space-time glides develops from within its own texture, but for that evolution of space-time to take place, the enactments must be endorsed by the vigil of the spectator. Irrespective of what design space-time follows, its commencement and conclusion would show up as dawning and termination in the plane of matter, and from *being* conscious to *becoming* the witnesser in the incalculable plane of the watcher.

In the progressive unfolding of the intellective capacities in the zoological hierarchy, the humans stay preeminently, abreast of the emotive nature and filled with the longing to understand the essence beyond the bounds of the usual. The attainment of such a status would echo a higher state of awareness, alongside, also announcing the awareness of the breadth that stretches beyond the neural perceptivities. Seizure of such a juncture—the awareness of human—would self-incite further expansion in the level of awareness. The higher the awareness, the larger the range of entities that are witnessed, the higher the separation from space-time through the flow of the world line, with respect to the particular observer.

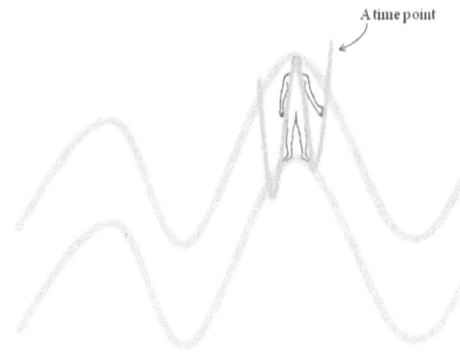

Figure 13.1. The full-length sweep of space-time in the window of witnesser that lodges side-by-side the embodiments of mass-energy spread. A time point in this compass would denote a symmetrical concurrence between the sensed and intellect, shown by the smaller symmetrical rendering.

Between being in a form of an entity to becoming the milieu beyond the entity lingers a fine line, utterly subtle and esoterically penetrating: the shadow of identity. And it is by the mere shift in the identification that the entity transmutes from entity to witnesser. The shift in the way one recognizes the self, element, or witnessing, then would occur through the annihilation of the elemental selves, gradually over the period of the life course.

The complete metamorphosis, through the exhaustive cracking of all of the elements, would then check off the end of cycle. The cessation also seals that no identification with any form of temporal mold remains. In like manner, the ending would also record the culmination in the act of witnessing, which prevailed only in reference to that particular set of elements. And yet, independent of that observer's view—purely as a part of the physical matrix—the amount of elements (the matter) that the observer lugged unvaryingly sustains.

Thus, in the utmost schematics, it is not that just the articulations of mass-energy stay in relativity; the collection of the physical elements itself flaps in relation to the over-drawn watcher. When nothing remains to be watched, even relativities are vanished, whether it is within the frame of materiality or between the matter and the witnesser.

The actualization of the empty space ($\{\ \ \}$, or $\{\varnothing\}$) cannot take place in the absence of the elements—the rest of the subsets in the power set. The empty space and subsets are commutual.

The Touch of Evolution

The uttermost flow that put on two distinctive shows—one of time, the other of identity—would exhale two scenes of evolution as well. In the floating parallel universes, the feats of beginning and end, although they appear cyclical, would be unidirectional in the evolution that occurs alongside the gradual transfer of identification from the elements to the observer; the route is irreversible and immutable. It is unidirectional because it involves stepping up and signifies the absence of stepping down. The unidirectional passage in the foremost view denotes successive treads in traversing that utterly deceptive thin line that shows up diffused between the arena of the matter and the ultimate of the watcher—the crossover that by definition involves the annulling of the bubbles of space-time. The gradual switch in the sense of identity would coincide with the abatement of space-time, bit by bit, culminating in, well, the point where nothing remains to be culminated. The passage is one way because the broken amalgams cannot reunite in relation to the observer; it is only by the realization of the observer, the identity shift, that the annulling is effectuated in the first place.

In the synergetic field of hyperspace, all possible events of the past are followed in the flowering of the current moment, the way we understand in the quantum field. And irrespective of what design is followed, two ingredients always cart the design: (1) the turning up of the route and (2) the mark of completion, both of which occur only for the observer and not in the structure of the purely physical universe. The click of cessation would be concomitant to the termination of the observer's raison d'être, as well, timelessly.

Conceptualizing Hierarchy

Within the weave of vitalities of the biological vista rests the image of "supernatural," much the same way the image of the being, instead of the stature of the witnesser, overshadows the perception of our own true identity. The idea of the supernatural perhaps stems in the shadow of the thought that the coming of life and the forces of universe are the casts of two unrelated phenomena. Upon acknowledging that the launch of creature is the fundamental part of the grand play, where the physical, subliminal, and witnesser parts of the universe integrate to effectuate an autonomous system, the deception of hierarchy in the spectacle of manifestations evanishes. With that realization comes the projection of the natural layout where space-time, by the ambit of the living system, becomes a self-administered fold under the umbrella of vigil.

On the other hand, the idea that the embodiments of space-time and the role of watching are inseparably amalgamated would again sprout only in the mirage of quasi-recognition; there isn't any contact, literally, that is cemented between these two utterly disjoined conducts. The nuts and bolts of physicality, for knowledge of the observer, and the mission of watching, for the journeying of space-time, are the two facets that remain in aloofness. The two do not merge; they only stand for each other. The shifting of the identification from being the entity to being divorced from the entity thus does not make any perturbative alteration in the thrust of the physical phenomenon.

The Observer in Structural Relativity

According to the recipe of $E = mc^2$, the higher the speed, the slower the clock ticks. The two-poles apart messages of this rationality are that (1) indeed, the higher the speed, the slower will be the ticking of time in relation to the observer, where the highest achievable speed is that of light; and (2) the other side of this very translation, which has not been acknowledged as much to be an issue of scholarly conundrum, is the inquiry of the tempo of time when the observer in essence abides outside the relative framework. The scientific argument on this puzzle is plain and simple: nothing stays out of the relative framework. Einstein's two theories of relativity staple this fact; accordingly, the observer that is seeing the event of movement in itself is also moving, albeit with different momentum. Therefore, in the relative structure, everything moves, just with different speeds, effectuating the differences in the ticktock, in the dynamism that carts the dots of relative times, but not a single unconditional one.

Keeping in view that time permeates only in dabs, administered by the dissimilarities in the journeying of the observer, and does not prevail holistically, how do we explain (1) the relativity of an observer against the cosmic frame that is expanding prodigiously as a holistic conformity, and (2) the rate of ticking for the traveler that rides the speed of light? Authentic or hypothetical, depending on the context, these two conundrums draw attention to a draft where an act of observation must carry through outside the fabric of relativity as well—if we are to ratiocinate the ongoing inflation of the universe as a single entirety. And in the plan where the observer hangs on apart from structural ambience, the second mystification can be dealt with as well. For the traveler that rides the speed of light, and thus reaches the edge of the relativistic framework, the tick of time ceases altogether and the movement or time becomes the game of manifestations. Retrospectively, the observer must catch the relativity of speed and time from outside of the texture for the arguments of relativities to take firm hold in their entirety.

The watcher that recognizes itself at the instance of self-realization does not flutter but remains manifested alongside dynamics, and it is through taking in the demeanor of the ultimate observer that we can quadrate all of the chief axioms into a concordant plasticity—whether in seeing the principles of physics concatenated in orchestration or in syndicating the bearings of the clairvoyant plane, by those very principles, into the transcendental framework. For that matter, the elements of the subconscious are the very same physicality that permeates the whole, and the emotive waves too are ephemeral in nature, passing in and out of the vista.

The mystification of how the poses of intelligent life emanate by the thrusts of universal laws has baffled the human mind since the beginning of time. The bewildering puzzlement of how biology conjoins the deeply interwoven principles has pervaded me too ever since I can consciously remember. The lucidity of the surfing emotions and feelings, and their impermanency, appearing and disappearing in relation to what is encountered externally, used to mystify me as strikingly as it does many of us. The structure that emerges from scientific knowledge in alliance to the way we perceive the reality evinces the connectedness of the cerebral and the mise-en-scène that meets the eye. And we all know that seizing the picture of the absolute composition takes far more than just cramming and assimilating the analytical outputs that by and large bring forward only distinctive shades of otherwise impenetrable texture.

The truth of witnesser that materializes in sentient selves cognitively, thus, must be further ratiocinated in the outcomes of observational research and the concepts emerging therein. The truest nature of reality, thus, can be interpreted by dovetailing two discrete sets of knowledge: (1) the span of the watcher that the seers point to, and which we all acknowledge from rationalizing the scientific inputs, and (2) what we gather from the field of empirical research. The glancing of both the scopes is a prerequisite if we are to not only ingest the uttermost nature of space-time but also absorb how biology and intelligence are woven within the same mise-en-scène.

The rest of this chapter picks up some of the other crucial subject matters such as the staunch symmetries of nature, the understanding of the big bang, and the phenomenon of black holes, from the perspective of the scheme that we have discussed. This may help accentuate the picture of existence we have tried to grasp, and further validate the descriptions of physics and mathematics in uncovering our true reality. We also have topics that utilize different tonalities in restating the schematic that we have been reasoning.

By the Symmetry

The symmetries space and time, time and movement, matter and antimatter, and straight-out and subliminal are not the only instances of duality that transpire in the dialogue of the whole. The shadow of symmetry imbues all facets of totality, and in many different manners—both in the print of concrete and by the medium of abstract. The whisk of symmetry endows the rendering of supplementarity and complementarity in the layout of the multiverse. The ping of mass to the smear of energy, the staple of enthalpy to the disentanglements of entropy, the demonstration of matter to the concealment of antimatter, the exactness of particle to the ambiguity of wave, the discharge of positive to the charge of negative, the stationing of space to the travel of time, the symmetries in macrocosm to the supersymmetries of hyperspace (stemming in the string theory), the debut of the beginning to the wane of the end, the transcendence of the curve to the measurability of the straight, the cast of elements to the background of empty space, the cinema of the universe to the implant of the spectator, the sensation of body to the thought of mind, the hologram of dream to the latch of world, the concept of human being to the concept of superbeing, and the two polarities of emotion are just a few of the endless ways in which the imprint of symmetry is seen in the regularity of nature.

A positron is an antiparticle of an electron, with same mass and opposite electric charge. An antiproton is an antiparticle of a proton, with the same mass as a proton, but carrying an exactly opposite charge. Veritable details on how partnerships invade quantum makeup are instructional and inviting—and I acknowledged many of these earlier. For our purpose here, it is adequate to recollect that the vanguard of scientific advancements emphasizes that the natural phenomena oscillate with flavors of symmetries.

We have also noted that the symmetries themselves materialize in rhythm, addressed as supersymmetries, in the signature of the M-theory—the most advanced view that currently holds the string theory, in addressing that matter and force are two separate views of the same blueprint. Whether we consider the empirical inputs or the clear-cut mathematical cues, the effective model of M-theory rests on the admittance of supersymmetry in the nature of symmetry.

The cohort in every partnership sprouts from within the tapestry of space-time, and ultimately the partnership of the witnesser and the witnessed also becomes an order of symmetry in the flow of the endless stretch of time.

Segregating Space-Time: Cutting off a Region from the Rest of the Cosmic Territory

In the landscape of space-time, there dwells a region that, although it dots the stratum of allness, hangs out disengaged, implanted with an unallied set of universal forces. Such a region, now addressed as a black hole, was first conceived by German astronomer Karl Schwarzschild in 1916, in the extrapolations of Einstein's theory of general relativity. The theory of general relativity calls attention to the causation of the bending of space-time owing to the force of gravity, and Schwarzschild realized that a spatial extent—the black hole—must take hold in the wake of increasing enormity of gravity over smaller and smaller spatial extents. The contortions due to the co-sweep of mass and gravity import that the higher density of mass, thus reflecting the strengthening of the gravity, will further pull in the matter, setting off a scene of perpetual caving in.

And when the crowdedness of mass attains a certain threshold value, the black hole ensues; the mass of the sun crammed into a spatial extent of about a two-mile radius, for example, would cause the befalling of a black hole. The forces of gravity become so intense that the mass aggregate within that spatial sphere starts to implode, effectuating the coming about of a center of gravitational pull, so enormously intense that nothing, within a proximity, escapes from its prodigious whisk to being devoured—not even a ray of light. And although the prevalence of the black hole was conceived in mathematical extrapolations, the accuracy of its existence has been thoroughly substantiated and readily archived.

On the basis of theoretic reviews and indirect experimental clues, it is now a commonly accepted fact that the whirl of a black hole agitates in the center of most galaxies; there is one in the center of the Milky Way, weighing about 3×10^6 times the mass of the sun—rather petite compared to supermassive black holes that are known to reside within the observed macrocosm. The largest identified supermassive bulk is about 6.6 billion times the mass of sun. It has been conceived that such a spatiality emerges when a massive star approaches the termination of its life cycle and the budding black hole continues to survive by devouring stars and the other cosmic matter that pulsates within the range of its gravitation, whereby some black holes swell into the supermassive ones.

Since anything that descends into the clasp of these feeding bodies gets lost forever, never to return—even an electromagnetic radiation—it is not possible to detect the occupancy of a black hole via a telescope. However, growing evidence has corroborated the occurrence of black holes merely by scrutinizing how a stellar entity is perturbed by the herculean swirl of suction as it approaches the scene that surrounds the ambit of the otherwise unseen black hole. Utilizing quantum approaches, studies led by Stephen Hawking, at the University of Cambridge, conclusively demonstrated that the territory around the event horizon—the effectuated boundary of the dominion of the black hole that permeates as an intersecting point of the two distinctive regions of gravity, one the black hole, and second everything else—exhibits radiation of visible and X-ray frequencies, causing a circling halo in the neighborhood of the black body. (The audience of *Interstellar* would be able to appreciate better the glowing display of event horizon.) The blackbody radiation, commonly known as Hawking radiation, is caused by the massive drag the immediacy encounters. Linking the manner in which the proximity behaves with the formulations of general relativity compellingly justifies, albeit indirectly, the presence of these enigmatical segregations, severed from the world that beats outside.

The induction of a black hole within a streak of continuum calls attention to the inception of a budding universe from within the depth of space-time. As soon as a black hole takes hold, the rest of the universe, with respect to that blackbody, comes to be a space-time of the past. The two continua emerge from the swing of one. And in the smooth flow of space-time, the coming of the ebb marked by the black hole, which lacks the spatial continuum with the rest, can be reconciled if we can value the transpiration of such articulation from the vantage point of the watcher. The two continua can be distinctively watched, but they cannot distinctively coexist when the forces between them are not shared.

The breaches and lacunae that sprout in the membrane-like discernment of space-time (figure 12.1) would then themselves manifest as spatial marks under the vigil of the accredited observer, the same way two discrete sets of universal forces do when a black hole takes hold.

Envisioning the Beginning: The Shadow of the Big Bang

The inception of the particle accelerator, some eighty years ago, provided direct access to viewing the fundaments of nature with crystal clarity. These utilities became an outright benefaction toward working out the theoretical views in particle and high-energy physics, including the attestation of antimatter and authentication of subatomic particles—the Higgs boson being the most recent, and the most talked-about, validation in the swarm of several others. The Large Hadron Collider, built by the European Organization for Nuclear Research, near Geneva, is the largest and the highest-energy particle accelerator, situated in an underground ring of tunnel stretched seventeen miles in circumference. It draws thousands of scientists from more than one hundred countries, bestowing ways to peer into the realities at the most fundamental levels. And considering that it possesses immense capacities to amass the matter into a miniscule amount of space, in the future, it may provide a way to glance into the mechanism of how black holes form and respire.

As the name suggests, the particle accelerator is utilized to accelerate particles, such as protons, with immense speeds as high as close to the speed of light. These particles at smashing speeds are then made to collide, in order to bring about the explosive eruption of a gargantuan proportion of energy, stuffed within a miniscule amount of space. It is hoped that the surge of this huge proportion of energy lumped up in such a small amount of space may cause the birth of a quantum scale black hole that is severely micro compared to the ones that live in the centers of galaxies, yet from the point of view of the fundamental principles, would be parallel to the cosmic counterpart nonetheless with regards to the mechanism of its birth, continuation, and death. Thus it is hoped that this approach may provide a further way to probe into the nature of the black hole, and scientists assume that such accessibilities may carve ways to see into the dimensions that seemingly remain warped and disconnected.

The collision of particles at enormous speeds is also conjectured to create a kind of ambience that, by the frames of physical laws, materialized at the instant of the big bang, and thus the particle accelerator can provide a window into all that transpired in the scene of the explosive eruption of the big bang—including the sway of parallel universes by the cosmetic of black hole.

The enactments of the two independent implementations of forces in the continuum of space-time, whether two sides of the event horizon or two spatial extents banded together via the conduit of a wormhole—a conceptualized tunneling between two far-flung regions of the universe, a sort of shortest route bypassing the usual road of space-time—cannot be reasoned until the gist of witnessing is accounted for. The emanation of the wormhole implies the flowering of a "new" route that itself is made up of the ingredients of space-time, which, although it seems to exhale from nowhere, factually might just be expelled as a region of space-time that is first disintegrated and then reconstructed, again only in reference to an observer, an event which within the arena of physicality did not certifiably occur.

Whether conceptualizing the commencement of the universe by the outcome of the big bang or assimilating the emergence of time from the compress of the gravitational singularity, the bearing of mass and the forces by which it pulsates cannot be taken as an average customary phenomenon. The gestures of hyperspace are ejected only in the discrete lenses of the observer, and thus either the beginning or the unfolding comes to be only a smaller order in the absolute field of higher dimension.

A Dollop of Asymmetry

Symmetry is the order by which the articulations in physics and mathematics are uttered. The taming in the order of symmetry to the perfect accuracy, though, would be detrimental to the transformation of space or the flow of time, for both features bank on the launchings of new manifestations entirely different from past ones. The order of the new can evenly flow only when the mode of symmetry is perpetually broken by every moment. In the theories that are built around the notions of big bang, it has been conceptualized that the universe started out in top-notch symmetry—both by the geometry of the physical arena and by the attribute of the four fundamental forces. But as time passed and things began to cool down, the plaster of asymmetry started to expel, and its occupancy continued to swell; the principles of the universe as well as the materialistic panorama bubbled with tinges of differentiations.

In the membrane of hyperspace, the ongoing manifestations of matter and the continuous pounding of time occur by the mode of symmetry breaking, whether pertaining to flow at the noticeable level or the expansion in the cognitive plane by the discrete windows of the watcher. The skewed symmetries pave the way into subsistence and vitality, and by that token they are the very tool that caused the commencement of the big bang or the unfolding of intelligence. Even the awe-inspiring symmetries that we relish in nature come with a dash of asymmetry fused in them—whether capturing the exhibits of zoological and botanical lives or the patterns of earthly and celestial zones. The flavor of asymmetry infiltrates between the materialization of matter and the flow of time. The seasoning of unevenness propels the streaming of the parallel universes and the ongoing side-by-side installations of the subliminal.

The prevailing asymmetry, the slight mismatch in every plane of significance, foremost typifies interactivity that the real plane must imbibe even to ensue, in the first place, and then to hang on thereafter. For without a nuance of lopsidedness, the universe, as well as the psychical, would plunge into the order of suspension and stagnation, plainly because progression entails the constant exhibits of differentiations fostering population shifts, which in turn execute the flow of space-time.

The dash of crookedness in the streaming of parallel planes is pivotal to the furtherance of conscious gliding, for the object becoming the measureless span of watcher entails observing the successions that are set up by the mixings of the worldly and subliminal planes, through the world line of life.

That Which Is Instinctive

In the subliminal realm, the swing of sensation and the wave of emotion emerge and submerge unfettered, and it is in these transitory attributes that we pick up the sense of self. As concrete as it is to the real world before us, the nature of self, for the most part, we have linked to the studies of occult and spirituality. And thus the general consensus is that in deciphering the truest nature, the reconciling of scientific facts with our own existence is irrelevant. It is as if there are two independent realities floating side-by-side, both having their own mini big bangs of completely different orders.

Like many among us, I was eager to drink in the subtleties of reality and alongside ratiocinate how the sentient presence becomes a part of the universal makeup. These quests drew me toward browsing some of the esoteric and spiritual accounts delivered by proficiencies such as those of Gurdjieff, Osho, and Krishnamurti, articulated in Tibetan-style write-ups and discoursed in the ancient Indian scriptures *Upanishads*.[2] I began intrinsically ingesting some of the pronunciations more seriously and zestfully than others. One morning while walking to my office at Hopkins, I blurted "Watch"—as perhaps a subconscious exertion to incite the docking of self-realization, a method about which I had read in the communications by Osho. And this time, out of the enormous number of ardent attempts that I had made time and again, on the spot and without the click of time, *it* took hold. There and then, without the permeation of time, the identification from self (or the elements of self) shifted to the span of the watcher that sweeps over and above "I" (or physical and subliminal elements). The cognitive urge to watch incited the perceptible latching of the watcher that remains an outmost identity of us all. And in that instance of cognizance, it's not that only the factor I was presented as an element, the all-encompassed relativity by which the "I" pounds, that is, all of the physical plane (subliminal included), was revealed as a single vista, rippling independently from the quality of pure witnessing. The scene that engages the marks of the innermost on par with the streaks of the cosmic plane, flashing a single compass of time-point, was snapped by the purity of witnessing, the only gear the watcher lugs.

Krishnamurti—Jiddu Krishnamurti, a writer and orator of philosophy and spirituality in the early 20th century—disseminates the absolute truth as being limitless, unconditioned, and unapproachable by any specific path, thus, is independent of religion and sect.[3] The ultimate identity can only be "realized," via the medium of the self. Human sentience seemingly is accompanied by an instinctive wisdom that there permeates a plane of surpassing dimension, over and above the worldly components, although viewed differently depending on the school of scientific or philosophical thought. Einstein related to cosmic religious feeling;[4] George Cantor saw it in absolute infinity;[5] Newton in the universe's law and order; Philosophers identify it in the nature of self. I heard an interesting and folksy point of view many years ago. It recounted that if all the knowledge concerning our elemental association with the universal reality, ascertained in scientific reasoning or assimilated from philosophical and theosophical dogmas, is kept deliberately hidden from a person, all along his or her childhood, the grown-up, regardless of the structural and worldly opinions that he or she lacked, will still emerge instilled with an impression that there exists a somewhat occult and far-reaching field over and above the conspicuous physicality. There seems to be just one kind of spatiotemporal design that warrants the inherent presence of such a sense unpretentiously ingrained in sentient selves. An arrangement where the being, by its very nature, invariably perches, in some measure, beyond the attributes of the self affirms that design.

In the fusion of space-time with the watcher, the evolution that occurs with respect to the watcher would follow a linear unfolding;

that is the only way to it. The cyclic attributes that over and over appear within the structural framework sooner or later cease to exist for the observer, bit by bit, as they undergo witnessing. Thus, in the unfolding of the absolute makeup, the decisive progression would entail the switching of identification—a straightway path that would involve transcending the ripples of the elements.

The Beingness in Being

The execution of a side-by-side display of conscious beings in the transcendental latticework of the multiverse can be envisioned by the help of a simple example. We noted earlier that the life path of a sentient being equates the coexpressions of inner self and worldly elements in the window of an observer and that the lineup of such windows engenders the surpassing flow of the higher dimensionality. In this multiverse flow of shared elements (the way we gathered earlier in chapter 9 on parallel universes, under the headings "Singling Out the Core of Nonduality from the Grand Frame of Duality" and "The Sweep of Consciousness in Parallel Arrangement"), the subliminal elements would be shared too. Thus a conscious being spontaneously comes to knit with another sentient form because they are in cahoots in the execution of the functional elements that are communal between them. The interactivity between the two aware beings accomplishes the execution of the functionality for which they stay in league. As a plain example, ejection of a creative element not only demands the longing for that creativity; it also needs the social setup to meet that ardor. Despite the fact that both the counter-sides of this creative element sojourn as a weave of space-time, the fulfillment of the creativity entails distinctive observing fenestrae of consciousness, where at the ultimate level the observer in and of itself stretches aside from space-time or the creativity.

The same synchronicity would hold in channeling the emotive aspect—be it a particular sentiment or a passion for a specific line of work. Though deeply tangled in the mesh of physicality, the emergence and the discharge of emotive attributes would take place in the lenses of observers—by the forces of aware entities in the give-and-take congregation. The get-together of give-and-take would surge as a functional blend in the dimensionless window of the witnesser (figure 13.2).

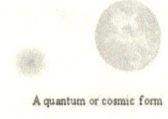

A quantum or cosmic form

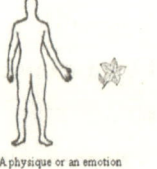

A physique or an emotion

An interaction

Figure 13.2. A spew of functionality in the arena of hyperspace.

Earth image credit: NASA Earth Observatory/NOAA/DOD

The emergence of the functionalities, their journey and completion, in the nature of conscious entities involves, at the most basic level, holographic projections of a donor and a recipient (as in the barber puzzle, in chapter 7; the ultimate of the barber is the watcher, and being the barber and not being the barber are functional elements)—an enactment that must be harnessed for the ultimate of the watcher to transpire.

The flow of conscious being would be akin to the channeling of conscious elements through the medium of beingness, in the view of the watcher. By that token, the sentient actuality doesn't stand for an embodiment that operates under the aegis of the watcher. It is rather that the watcher by and of itself cruises through the cast of the world line (the life cycle), where every throb of time churns by the docking of a being—a quasi-state of actual beingness. Thus, at the ultimate level, witnessing of space-time through the bay window of the watcher becomes a rather unavoidable enterprise—if the consciousness were to continue unfurling.

Pedantic Interlinkages

With the towering progress in the way we understand nature and its cardinal symbolizations, none of the specific academic fields on its own can furnish a full-length chart of the universe. The essence of mathematics can't be fully appreciated without an ethereal inclination for philosophy; the subjects of biology and medicine can't be infiltrated deeply enough unless there is a heavy bent for physics and chemistry in seeing all the fine details. The fields of mathematics and physics mount so deeply interwoven that one niftily echoes the other; to say the least, the shimmer of mathematics perches the perspicuous forms that physics delivers.

Toward rationalizing the spectacle of totality, the demeanors of the living arena must be interlaced with the canons of the physical sciences; the role of consciousness can be espied only by doing so. And to pin down the reality of the enigmatic mind and the concurrency of the swaying universe, we need to braid the rendering of mathematics with the delineations in physics so that the sphere that nests beyond the neural handles can be espied as well. We would

need to capture the whole order even to appreciate the purely experimental facts or sense the meaning they convey.

Delineating the Abstruse

The nonduality in the order of the hyperspace can also be delin-
eated in the commentary of the polynomials. The established connection between the circumnavigations of mental and direct planes
is resounded in the delivery of polynomial expression. Figure 13.3
illuminates the surmise of nonduality, where the holographic demeanors of internal and external are shined by the materialization
of a quartic function. The pairing of the cognition and that which
strikes the eye would occur for every time point (as I mentioned
earlier, the matching shouldn't be taken to be literal, but rather in a
way that involves psychological bent), and so the quartic function
here emblematizes the click of a moment. The function continuously flows, ebbs, and readjusts as time continues to grow into the
abysmal of the interminable boundlessness.

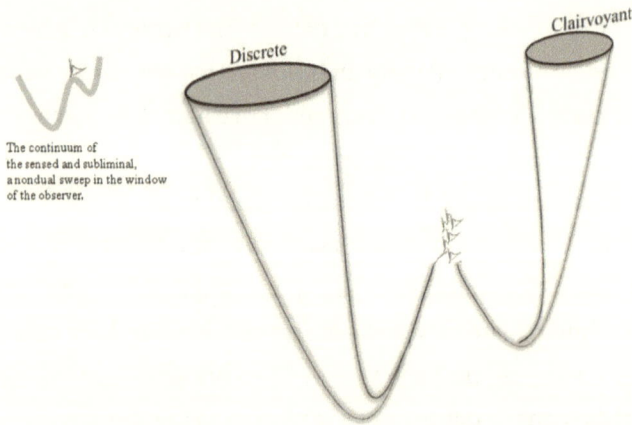

Figure 13.3. The deceptive swelling of space-time and the discharge of parallel universes, by the launch of multiple espying, of the very same sensed and subliminal attributes. In the view of the utmost identity watcher, the universal functionalities continue and evolve as a single interactive nondual framework.

In this facile gauge of the polynomial, the neural apparatus lies on the axis of symmetry, with the level of electromagnetic potentials that the eye reads on one side and the level that involves personal bent on the other, both shaping the view of the watcher (figure 13.3).

The overlapping collage of the parallel field would then come to play by the cast of more than one set of neural gadgets planted at the axis of symmetry, instead of just one, which indicates a single biological system (figure 13.3). All the elements of physical and mental makeup would remain in a nondual state of the universal pool in the parallel field. For instance, the bumping of two sentient entities into identical emotive reflexes would not signify the occurrence of two reflexes; it would only insinuate that the particular reaction, although picked up by two beings, permeates singularly in the supreme weave of space-time (figures 9.1, 13.3).

The untamed shades of mathematics and physics spotlight the portraiture of the ultimate unification, of all that there is, when the constituents of the subliminal terrain are also conjectured as a show of discrete entities, buoyant in the fluidic amplitude of the utmost continuum.

Noms de guerre that whisk laterally, aligned in the textile of the tangible, smear to share the parcels of space-time in the protean field that hangs about harmonized with the direct vista, engendering a scope that pulsates ad infinitum.

The debarkation of the beholder is synchronous to the exposure of that which is beheld.

The mise-en-scène of reality that encompasses ingredients of consciousness and physicality permeates in a single overarching script, partly visible in the window of eye, and fully captured in the view of the watcher, for every tick of time.

No sooner does the engraving of a universe clock in than the actualization of parallel universes takes the stage.

In the estate of truest reality, the "order of whole" is reflected in every gaugeable display, for any scene from the vantage point of the watcher is the shimmer of the whole piece. That would be why the spatial ounce of quantum size mirrors the phenomenon of the gigantic cosmic universe. In the window of the measureless witnesser, the vast bound of cosmic frame and the infinitesimal compass of the quantum latch identically.

BRIEF NOTES

DISCONTINUOUS CONTINUA

a.   Aesthetics or Mathematics

In the pictorial sense, the accent of mathematics doesn't just reflect a disposition of a single academic subject. Labeling mathematics as a separate straight-out branch seems to imply that it is a constituent of a larger academic curriculum. Apart from its compulsory presence in all fields of knowledge, application and creativity, the tone of mathematics, by and of itself, pitches the blueprint of the ultimate landscape in utter purity.

The unadulterated recitation of the reality that mathematics whispers springs from a subtle and constitutional attribute. Any prescript in the rubric of mathematics comes with an image of the empty space, lying abaft the numerals. The mathematical articulations as entireties cease to prevail without the bed of the background medium in which they are smoothly structured. And it is for this portrayal of the integrated whole that even the most abstruse bearings of reality can be duplicated in forms of discrete numerical embodiments.

The aesthetical shine, the touch of truth and the power to illuminate, transpires by the mode of mathematics because (1) the numerical declarations come into view through their own beyond and (2) they exist only as a systematic whole, in different notational ways. Thus, in the elucidation of the orderly exclusives, the numeral fields impart an archetype of the ultimate nature—the ultimate that subsumes the fundamentals of matter and sentience, and the mechanism by which the two intermix into a single mold.

The structure of the world, or that of the inner plane in the view of the watcher, is not mathematical; rather it is that the inflection of mathematics and the pronunciation of the universe mirror one and the same conformation. The materialization of mathematics takes hold as soon as we start to evaluate.

b.   Chaos Is a Misnomer

Fractals are structural motifs built by the iterative tracing of a specific geometrical contour; the truly appealing iterativeness can be concocted utilizing mathematical formulations. The imprints of fractal framework occur ubiquitously in all facets of nature and the universe—from anatomy, growth, and development in the physiological arena to its reflections in the figuration of clouds, snowflakes, crystals, lightning, weather patterns, and mountain ranges, to the presence of fractal demeanors in cosmic sculptures, and in the spatial distribution of the universe as a whole.

The term "fractal," coined by a French American mathematician, Benoit Mandelbrot, is derived from a Latin word, *"fractus,"* which means "fractured." A fractal can be fractured into an array of a particular outline, gradually differing in size, where each one of the silhouettes images the mold of the whole. The enunciation of the fractal elucidates the show of self-symmetry. The beauty of the fractal can be appreciated in mathematical depictions, and some online sites provide pictorial delineations, featuring ravishing details in exquisite complexities. In some instances, such as the lineament of lightning or the spatial distribution of the universe, the show of a fractal at first sight may appear chaotic, lacking the presence of any discernible motif. However, these apparently chaotic portrayals de facto come in sharp mathematical accuracies.[1]

Along the same line, the thermodynamic component of entropy too is an essence of the functioning of the universe, plainly because its expression controls switching between the quantifiable states—a necessity for transformation to occur and time to flux. Thus the element of entropy cannot be quantified, for it reflects the haphazard—a crux to bring the states of transition into the nature of space-time. And by that token, the breaking of enthalpy channels a quantitative value to the following entropy—the correspondence that in turn allows the manifestation of new forms, developments, and changes.

There remains an order in chaos. Whether seen in fractals or in entropies, the feature of chaos reflects the brilliance of transcendence that shepherds every route by which reality travels.

In Albert Einstein's words,

> "You believe in the God who plays dice, and I in complete law and order in a world which objectively exists, and which I, in a wildly speculative way, am trying to capture."

Ultimately, the tossing of the dice would also be a part of the utmost order.

c.   Science and The Rest

The methodic study of the observed and discerned facts of the physical world is the definition of science, in its simplest form. The information that stems from the fields of biology, chemistry, and physics comprises discrete pieces of erudition that are pieced together to depict the whole via the predilections that reside in the subliminal faculty. In envisioning the fold of the whole, the averments of mind, thus, become a body of facts too—as much as those that the mind detects.

The theosophical concepts of Ein-Sof from the Kabbalah or Brahman and Atman from the Upanishads are conceivably based on what was recognized abaft the derivatives of the mental plane. The portrait that science yields when interweaved with that which is perceived above the cerebral elements can offer an all-inclusive draft of space-time. The piecing together of the two, science and the perceived, is utterly crucial if we are to understand the utmost texture with pedagogic fervor. Not to mention that the reality at the absolute level can be inferred only in totality, when all the residing spheres of the megacosm are credited.

The overarching realm that is beheld, enmeshed with the dots of direct plane, reconciles with the elaborateness of space-time that science disseminates.

The fact that the experimental verities and the discernments that lie beyond the borderline of science can be depicted in the language of mathematics is yet another alluring matter.

d.   They Are Not Two Things

Advaita Vedanta is considered the most advanced subschool of Vedanta, the ancient Hindu scriptures dictated to disseminate the philosophical interpretations on the nature of the ultimate reality that one perceives at the juncture of self-realization. Originated in Sanskrit, the word *"Vedanta"* translates as "finality of the purpose" or "end of the goal." The preceding word, *"Advaita"* is a sum of two words: *"a,"* implying "not," and *"dvaita,"* meaning "two." The communication of Advaita Vedanta professes the conduct of nonduality in the uttermost realm of reality. And although the philosophical text refers to the declaration of nonduality at the metaphysical level, we have acknowledged previously that the essence of nonduality would be intrinsic to the system of physical entities as well.

For a façade of an entity, of any form—elements of matter, attributes of mind, or the whole universe in totality—the verity of existence equates to the truthfulness of coexistence. And at the level of the physical panorama, the pitch of coexistence translates into an illusionary shadow that there subsists more than one.

The Calabi-Yau manifold is a mathematical model that can metaphrase the multidimensional space-time fold. In the tone of numbers, such a space is characterized by the assertion of non-vanishing complex fields—a bound that remains nonzero everywhere, bringing about the order of the numerical continuum. The multidimensional circuit must proclaim a clean sweep of a single stroke raised in continuity, by which the circuit can be represented as a holistic span in the streaming of space-time. Whether it is the ambit of multifarious dimensions or the array of parallel universes, the scientific intonations, too, impart that the physicality breathes in nonduality, the same way philosophical doctrines affirm for the state of the nonphysical, or beyond physical.

$$H^k(M) \cong H_{n-k}(M)$$

The equation announces the nature of duality and phrases a rendition of symmetry that embeds the string theory. I noted in chapter 9 that the deduction averred by Henri Poincaré utters that for an n-dimensional closed manifold M, the kth cohomology group H^k is isomorphic (i.e., structurally identical) to the $(n - k)$th homology group H_{n-k}, conveying the beat of duality—that there are two ways to communicate the same tonality. And in the panorama of reality, such diction whispers that for each topological manifestation there transpires a structurally identical cohort. The avant-garde ratiocinations in the field of physics, including the string theory, that sleekly intimate the quintessential of the unified scheme axiomatically accommodate the stroke of duality by every crumb of their formulation.

Irrefutable as it plays within the fabric of the physical sheet, from above the outskirts of space-time the ignis fatuus of duality dissolves by the convergence of the image and the coimage into a single fold of a nondual state.

e.   Entity: Discrete or Abstract?

$$\hat{f}(\xi)=\int_{-\infty}^{\infty}f(x)e^{-2\pi ix\xi}dx$$

$$e^{2\pi i\theta}=\cos 2\pi\theta + i\sin 2\pi\theta$$

The first equation, known as Fourier transform, decomposes a function into its oscillatory functions. This operation is extensively used in the many fields of the physical sciences, especially in engineering and mathematical physics.

The second equation, known as Euler's formula, illustrates the wave function in a complex exponential field. The formula pervasively appears in the branches of mathematics and physics.

Although the cadence of mathematics is recited by the pitch of discrete and explicit, the numerical formulations, nonetheless, can adeptly lock up brilliant modulations toward broadcasting abstractness. Wherefore the labyrinthine physical conformities can be drafted and alongside grasped.

The numerically pronounced topological object Calabi-Yau manifold acts as a beau ideal toward understanding the magnum opus of the theory of general relativity, and this manifold surfaces wherever the envisioning of the higher-dimensional continuum paves the way toward explaining away the advanced findings.

f. In Words

The incidences that deeply touch the inner emotive core are relatively scarce to come by, and we all long for a rendezvous that can stir a meaningful difference. The journey of life is routed alongside the discovery of the self, but often we remain oblivious to the significance of what lies behind words, and perhaps the sights. Thereupon, as we live through the scenes of time, the layers of memories engraved with vivid impressions continue to pile up as a sealed amassment of discrete time slots. Every incidence plays a role; only some have the thrust to take one away from the ropes of time and the gravitation of space.

About the time I was finishing the preliminary draft of the present write-up, I chanced upon a complaisant and an exquisite movie titled *Hearts in Atlantis*, the storyline of which was taken from a fictional book by the same title, narrated by a notably popular author, Stephen King. The movie was released in 2001, but I did not become aware of it until just a short while ago. The movie surely has a sublime and aesthetic appeal to it. However, how deeply one is touched by its tonality would largely depend on the bent of the emotive impression one already harbors. As I watched this movie, I was recalled into a cranny of time that had remained frozen within me for about twenty-five years. The alcove of time remained wrapped so effectively that all this while, within my field of awareness, it was as if it had never existed. But upon the remembrance, the hidden subconscious surfaced utterly powerfully, lucidly crystal and robustly emotive. The reliving of the nostalgic recollection was far more swaying and real than the feeling that was cognized in the living of the actual time so many years ago.

I would not write about the instance mainly for its enormity and value; but to quote the least, the recollection of this submerged

past in itself carries profound meaning, mainly for its precious-
ness that remained deeply ingrained in me, untouched for so many
years, waiting to be acknowledged.

One may say that the watching of the movie and the recollection of the personal memory overlap in time. I would say they
are two sides of the same utterance. Despite there being stark and
consequential differences between the story line of the film and
the reality of my experience, from where I stand, both portray interchangeable meanings.

ACKNOWLEDGMENTS

Recognitions cannot be formalized hands down, for in conjunction with everyone who played a role, all that comes to pass pitches an ingredient into that which continues to become. Still, a few names should be duly mentioned, especially the ones that are directly connected to the mechanics of this draft. Slots of time that directly connect the span in which this book is written also warrant acknowledgment.

Much of this communication appertains to my own interpretations of the observed outputs, springing in the foremost fields of physics, and viewpoints on the compelling concepts, voiced in the vivid expressions of mathematics—in a pursuit to picture the core of the ultimate space-time through the forces of scientific revelation. With the anticipation that it would surely help further sharpen the sketches I present here, I approached a selective group of the foremost academicians that currently dominate the advanced fields of mathematical physics, who I believed might find interest in the formulations that are accounted here, to solicit their feedback on specifics. Some of those who responded refused.

This actually helped in many ways, for then I had no choice but to turn to additional rounds of ponderings and assimilations in ascertaining that the scientific ratiocinations that are inserted in the schematic presented here are drilled justly and employed boldly. In the course of time, after revamps, tune-ups, and grindings, all the bits and pieces that formerly seemed to be floating aloofly started to articulate into an all-in-one piece of coherency.

From where I stand, the avant-garde uncovering in the field of physics and intricate phrasings in the complexion of mathematics—although I utterly relish them and have been reading about them for a few years now—are relatively new areas of study. My field of research for many years, until very recently, has been in areas of biophysics, specifically thermodynamic approaches to understanding the protein structure–function relation—which although in many ways falls within the boundaries of the physical sciences, when it comes to utilizing scientific deductions in interpreting the ultimate motif of space-time, is far removed from the route that could potentially lead to the ultimate scene. While the current draft does reflect instructional feedback from some noted academicians, I am open to additional peer review glances by a specialist eye from the physical sciences.

As for the neighboring space-time turfs that are directly connected to the docking of this correspondence, there are primarily two compasses. The first one is the span of ethereal and soothing aloneness that rendezvoused for three years in Baltimore, where though I was engaged to a research work of biomedical relevance I also had academic freedom to assimilate and explore other expository curricula that were openly a part of the university structure at Johns Hopkins University, where I worked in the capacity of an associate research scientist. I was associated with the group led by E. Freire, and along with the ways and insights of structural thermodynamics, I enjoyed reconciling theoretical values with experimental observations. The interactivity of the group, and the theory and experiment, felt scientifically efficient and personally pleasing.

My time at Hopkins also benefited me in a different way. There the academic activities were interspersed with a vastness of ad lib reads and extended walks, or just periods of solitude during the evenings and over most of the weekends; the breadth provided the essential ticking in the subliminal bounds of my own existence.

The second stretch of solitude manifested in the convenient basement of Sandeep's house, where I moved after I let protein science go from my universal landscape, again for almost three years, and where the misty mold of this communication started to appear.

I want to thank Bernhardt Trout, from MIT; Yuk Sham, from the University of Minnesota; and Ernesto Freire for agreeing to review this manuscript. Yuk Sham provided a great deal of practical advice on how to bring this interdisciplinary subject to the market effectively and make the meanings better approachable. Bernhard Trout is warmly thanked for his most sincere input on the subject overall, for pointing out potential loopholes, and for bringing to my attention some of the crucial arguments missed. His suggestions helped enormously in making the messages of this communication crisper and firmer. I have always deeply valued the ongoing support of S. S. Bisht in everything I attempt, and I truly recognize his encouragement on the topic of my current interest.

In connection with the individuals who directly helped in channeling the literary mechanics of this commentary, I want to thank the editors Hillel Black and Pamela Contractor for providing detailed comments and suggestions on the preliminary manuscript. The editorial expositions that came from Pamela at first glance appeared like brutal doles of totally unnecessary grumblings. Much later, while way into thorough rectifications did I realize that they were much deserved. I want to acknowledge Claudia Lipschultz for her friendly notes on writing style and her suggestions on improving on the firmness that mobilized the rewriting with firmer scientific reverberations. Holly Monteith is thanked for a thorough job on copyediting the final draft, and for her valuable comments and suggestions. The Archway team is acknowledged for their editorial work and valuable inputs in finalizing the manuscript.

I want to acknowledge Sandeep and Manushi (I am married to Sandeep, and Manushi is our daughter) for their ongoing support and the affable company that they are—a cardinal ingredient that was crucial for me to see this writing through.

And there certainly are others who played positive roles, but it won't be possible to mention each of them here. I recognize their dispositions.

GLOSSARY

Algebra

A broad field of mathematics with many subsidiary branches that, in a nutshell, establishes statements of numerical interrelationships, from plain arithmetical formulations to the structure of the complex fields. The equation $x + 4 = 8$ is an example of a very simple algebraic equation, where the value of the unknown quantity x is stated by an expression.

Boundaryless

In the context of the absolute space-time, the boundaryless nature points to the presence of a span alongside the four-dimensional boundaries of space-time. In the current context of mathematical physics, "the boundaryless" would also refer to the boundarylessness of space-time, where the apparent boundaries remain warped. Thus there isn't any actual boundary present, as the universal structure remains closed and continuous.

Calabi-Yau manifold

Named after mathematicians E. Calabi and S.T. Tau, the Calabi-Yau manifold is a multidimensional mathematical form that is projected utilizing complex algebraic equations. The manifold serves as a model in understanding current physics views, such as the existence of extra-dimensional space-time and the shade of mirror symmetry in the string theory.

Calculus

A branch of mathematics that is designed to identify the rate of change for a given process. Calculus comprises two subbranches: differential and integral. While differential calculus deals with ascertaining derivatives, integral calculus concerns finding integrals in delineating the rate of change.

Cantor

Georg Cantor (1845–1918) was a German mathematician and the founder of the set theory. He is known for his spirited quest to understand the ultimate nature via discerning the disposition of infinity.

Cardinality

This term, used in set theory and introduced by Georg Cantor, points to the number of elements present in a set. The feature of cardinality is especially valuable in understanding the nature of infinite sets and their power sets.

Cartesian plane

Named after the French mathematician René Descartes, the Cartesian plane is a coordinate field that has negative sides, along with positive ones, for both x and y axes. Such a graphical tool is of indispensable utility in many branches of advanced mathematics and mathematical physics.

Concurrentness

Concurrentness refers to the state of side-by-side occurrences. In this text, it indicates the cast of duality in the materialization of the universe. Space-time, particle-wave, mass-energy, time-movement, enthalpy-entropy, and the streaming of parallel universes are some of the obvious examples.

Curve
    Curve can be defined as a constant curvature that occurs with respect to the straight. In the context of this book, curve is a property that comes to exist only in relativity; the appearance of curve is a signature of structural relativity.

Dark energy
    About 70 percent of the entire mass-energy of the universe, dark energy is contemplated to be the driving force behind the constant expansion of the universe.

Dark matter
    About 23 percent of total matter is deduced to be dark matter. Dark matter remains inert to electromagnetic radiation, and because of this, its presence cannot be established telescopically. The habitation of dark matter is considered based on the gravitational influence that it has on the visible arena.

Dimensionlessness
    Dimensionlessness speaks of an actual (not metaphorical) order in the flow of the integrated space-time.

DNA
    DNA (an abbreviation for deoxyribonucleic acid) is a biomolecule that the carries genetic code in all living organisms, except in the case of RNA viruses. It is a helical structure made up of two antiparallel polymeric strands.

Duality

In the branches of mathematics and physics, there are many different ways the conception of duality is recognized. Among them, the Poincaré duality relates closest to the topic of this book. Poincaré duality suggests that a structure with a particular set of physical properties can exist in two different ways. Poincaré duality also sketches out the imprint of symmetry and supersymmetry in the advanced forms of string theory.

$E = mc^2$ (where E = energy, m = mass, and c = the speed of light)

This equation was devised by Albert Einstein in the year 1905, calling attention to the equivalency of mass and energy. The equation is core to the theory of special relativity (suggested in 1905) and subsequently to the theory of general relativity (offered in 1916)—both by Einstein.

Enthalpy

Enthalpy is a thermodynamic energy that refers to the favorable electrostatic interactions for a given physical system.

Entropy

Entropy is a thermodynamic energy that is the opposite of enthalpy, pointing toward the natural tendency to acquire freedom and maximum flexibility. Thermodynamically, a physical process is driven by compensations between enthalpy and entropy, as both enthalpy and entropy play important roles.

Euclidean geometry

A mathematical system laid out by a Greek mathematician, Euclid, in his book *The Elements* (ca. 300 B.C.), addressing the fundamental axioms of plane and solid geometry, simply known as Euclidian geometry. It is a standard geometry that appears in textbooks.

Event horizon
    A periphery that segregates the external space from the immense gravitational pull of a black hole. The compass of an event horizon in effect defines the inhabitance of a black hole.

Fibonacci sequencec
    A sequence declared by an Italian mathematician, Leonardo Pisano Bigollo, commonly known as Fibonacci:

$$0, 1, 1, 2, 3, 5, 8, 13, 21, 34, 55, 89, 144, \ldots$$

The sequence follows an underlying pattern. Beginning from the second place (i.e., 1), every subsequent number is the sum of the previous two. The Fibonacci sequence reflects the golden ratio (ϕ) in its unfolding. See chapter 3 for their interrelationship.

Function (mathematics)
    Function establishes the correspondence between input and output, defined for a given problem. For instance, $f(x) = 10x$ is a very simple form of a function that says that the input must multiply by 10 to give the output.

Gurdjieff
    George Gurdjieff (1866–1949) was a leading Russian spiritual teacher who ministered ways to awaken consciousness—commonly known as the fourth way.

Hologram (current physics)
    A hologram is a three-dimensional projection of the extent

that in and of itself resides as a two-dimensional plane. Toward accommodating gravity within the structure of the string theory, scientists now predict that the three-dimensional universe that we see exists only by way of holographic impressions.

Homology (mathematics)
The custom of homology (from the Greek word "*homos*," meaning "identical") is utilized in advanced algebra and concerns mapping groups and complex topological forms.

Irrational number
An irrational number exhibits unlimited decimal expansion that lacks an orderly pattern.

Isomorphism (mathematics)
Isomorphism (from the Greek words "*isos*," meaning "equal," and "*morphe*," meaning "shape") notates the structural match between mathematical groups in reference to the interelemental relationships that they carry. The mapping is routinely utilized in the branches of abstract mathematics.

Manifold
In the context of mathematical physics, a manifold is a topological structure that exhibits dimensional extents higher than the familiar ones that we perceive in the universe. The objects are of enormous utility in understanding current concepts and discoveries.

Möbius
August Ferdinand Möbius (1790–1868) was a German mathematician and astronomer who is most commonly known for his viewpoint of a topological structure, the Möbius strip, a

nonorientable geometrical shape with only one surface. The name of the mathematical technique "Möbius transformation" was given in his recognition. The topography of the Möbius strip was also envisioned by another German mathematician, Johann Benedict Listing, around the same time August Ferdinand Möbius charted this spatial form.

Non-Euclidean geometry

Evolved from the axioms of Euclidian geometry, non-Euclidean geometry charts advanced forms of spatial descriptions—such as elliptical and hyperbolic spaces—that are routinely utilized in modern formulations, for example, in the theories of relativities and descriptions of manifold and hyperspace.

Osho

Osho (1931–90) was an enlightened Indian mystic, philosopher, spiritual leader, and public speaker who traveled across the world disseminating ancient philosophical teachings. He is mostly known for interpreting ancient scriptures and conveying the meaning behind spirituality and human freedom in a nontraditional way.

Periodic boundary condition

A theoretical premise utilized in computational simulations to prevent artifacts that can occur as a result of the restraints caused by the presence of an actual boundary. In the periodic boundary condition, the boundaries of the system remain in continuum.

Pi (π)

Initiating as 3.141592, pi (π) is a transcendental number defined by the ratio of the circle's circumference to its diameter.

Quantum (physics)

The quantum is the smallest amount of mass-energy that can subsist as a discrete entity. The quantum of light is known as a photon.

Quantum mechanics

Quantum mechanics is a branch of physics that involves looking into mass-energy behavior at microscopic scales, where the dynamics are of the order of Planck's constant, approximately 6.626×10^{-34} J·s.

Set theory

The set theory is a branch of mathematics established by German mathematicians Georg Cantor and Richard Dedekind (in 1870) that outlines the sets of finite and infinite objects, their properties, and their interrelationships and continua.

Space-time continuum

Space and time interfuse for the nature of physicality to manifest and forge ahead.

Space-warp

Space-warp refers to the continuum of space that occurs because the spatial ambits stay closed onto themselves. Thus, in the mold of space-warp, what appears to be an infinite extension is just a finite sphere that is shaped by the implant of the continuum, through all of its dimensions.

Supersymmetry (mathematics and physics)

The conception that thrives in modern theories of unification at a very basic level suggests that the symmetries in space-time are accompanied by supersymmetries, where matter and force too pairs.

Transcendental numbers

Transcendental numbers are those irrational numbers that are nonalgebraic. For example, π and e are transcendental numbers, whereas $\sqrt{2}$ is just an irrational number, as it can be expressed algebraically; $\sqrt{2}$ is algebraic because it can be a root of a polynomial expression. For example, x in the expression $x^2 - 2$ is $\sqrt{2}$, or $\sqrt{2} \times \sqrt{2}$ gives 2.

Trigonometry

From the Greek words "*trigonon*" and "*metron*," refereeing to the measuring of triangle, trigonometry is a branch of mathematics that involves studying the properties of triangles, for the relative nature of their sides and the corresponding angles between them. The defined trigonometric functions are utilized to delineate cyclical phenomena.

Vitruvian man

Vitruvian man is a celebrated illustration made by Leonardo da Vinci. The sketch accompanies notes based on works by a well-known architect, Vitruvius. The drawing indicates the layout of natural proportions in reference to geometrical viewpoints.

Wormhole

Appearing only in theoretical purviews, the idea of a wormhole is indicative of an appearance of a bypass tunnel of space-time that would connect two far-off regions of cosmos.

NOTES, BIBLIOGRAPHY, AND REFERENCE TITLES

Chapter 1
1. Jacob D. Bekenstein, "Holographic Universe" Scientific American Reports (2007) p.67.
2. Mario Livio, *Is God a Mathematician* (New York: Simon and Schuster, 2010).
3. Mario Livio, *The Golden Ratio* (New York: Broadway Books, 2002).
4. William Dunham, *The Mathematical Universe: An Alphabetical Journey Through the Great Proofs, Problems, and Personalities* (New York: John Wiley & Sons, Inc., 1994).
5. Edward Rothstein, *Emblems of Mind: The Inner Life of Music and Mathematics* (Chicago: University of Chicago Press, 2006).
6. Ian Stewart, *Why Beauty is Truth: A History of Symmetry* (New York: Basic Books, 2007).
7. Brian Greene, *The Elegant Universe* (New York: W. W. Norton and Company, Inc., 2003).
8. Stephen Hawking and Leonard Mlodinow, *The Grand Design* (New York: Bantam Books, 2010).

Chapter 2

1. Amir D. Aczel, *The Mystery of the Aleph: Mathematics, the Kabbalah, and the Search for Infinity* (New York: Washington Square Press, 2000).

2. The Mathematical Universe Hypothesis of Max Tegmark (reference articles):

    Max Tegmark, "The Mathematical Universe," *Foundations of Physics* 38 (2008): 101–150.

    Max Tegmark, "Mathematical Cosmos: Reality by Numbers," *New Scientist* 195, 2621 (2007): 38–41.

    Max Tegmark, *Shut Up and Calculate*, September 25, 2007, tinyurl.com/6pjjxp.

3. William Dunham, *The Mathematical Universe: An Alphabetical Journey Through the Great Proofs, Problems, and Personalities* (New York: John Wiley & Sons, Inc., 1994).

4. Roger Penrose, *The Road to Reality: A Complete Guide to the Laws of the Universe* (New York: Vintage Books, 2004).

5. Amir D. Aczel, *Fermat's Last Theorem* (New York: Basic Books, 1996).

6. Simon Singh, *Fermat's Enigma* (New York: Anchor Books, 1997).

Chapter 3

1. Alexander J. Yee and Shigeru Kondo, "Round 2 … 10 Trillion Digits of Pi," last revised December 28, 2013, http://www.numberworld.org/misc_runs/pi-10t/details.html.

2. Peter Alfeld, "Pi to 10,000 Digits," University of Utah, August 16, 1996, www.math.utah.edu/~alfeld/math/pi.html.

3. Ron Knott, "Fibonacci Numbers and the Golden Section," University of Surrey, accessed December 10, 2014, www.maths.surrey.ac.wk/hosted-sites/R.Knott/Fibonacci/fib.html.

4. Coldea, et al., "Quantum criticality in an ising chain: Experimental evidence for emergent E_8 symmetry," *Science* 327 (2010): 177–180.

5. Cristóbal Vila, "Nature by numbers: The elegance of mathematics meets the breathtaking complexity of the natural world," *AEON-VIDEO*, http://aeon.co/video/science/nature-by-numbers-a-short-film-about-the-elegance-of-mathematics/.

6. H.E. Huntley, *The Divine Proportion: A Study in Mathematical Beauty* (New York: Dover Publications, 1970).

7. Ron Knott, "Fractions–Decimals Calculator Version 2.6," University of Surrey, last updated September 10, 2014, www.maths.surrey.ac.uk/hosted-sites/R.Knott/Fractions/FractionsCalc.html.

Chapter 4

1. David Joyce, "Euclid's Elements," Clark University, accessed December 20, 2014, aleph0.clarku.edu/~djoyce/java/elements/Euclid.html.

2. University of St. Andrews, "Famous Curves Index," November 2014, www-groups.dcs.st-and.ac.uk/~history/Curves/Curves.html.

3. On heavily curled up spatial dimensions, appearing in the scientific interpretations of reality:

Kaluza-Klein theory: The information can be easily accessed online. A site for a general idea is superstring-theory.com/experm/exper51.html.

For overall background and how the idea of curled-up dimensions fitted the paradigm of string theory, see the following titles:

Brian Greene, *The Elegant Universe* (New York: W. W. Norton and Company, Inc., 2003).

Roger Penrose, *The Road to Reality: A Complete Guide to the Laws of the Universe* (New York: Vintage Books, 2004).

4. Clifford A. Pickover, *The Math Book: From Pythagoras to the 57th Dimension, 250 Milestones of Mathematics* (New York: Sterling, 2009).

5. Richard Padovan, *Proportion: Science. Philosophy, Architecture* (Abingdon: Taylor & Francis, 1999). (See Adolph Zeising's work in chapter fifteen.)

6. Kimberly Elam, *Geometry of Design* (New York: Princeton Architectural Press, 2001).

Chapter 5

1. John Derbyshire, *Unknown Quantity: A Real and Imaginary History of Algebra* (New York: Plume, 2007).

2. Charles Seife, *Zero. The Biography of a Dangerous Idea* (New York: Penguin Books, 2000).

3. Keith Devlin, *The Language of Mathematics: Making the Invisible Visible* (New York: Holt Paperbacks, 2000).

4. The "Mathematical Universe Hypothesis" of Max Tegmark: Max Tegmark, "The Mathematical Universe," *Foundations of Physics* 38 (2008): 101–150.

Max Tegmark, "Mathematical Cosmos: Reality by Numbers," *New Scientist* 195, 2621 (2007): 38–41. Max Tegmark, *Shut Up and Calculate*, September 25, 2007, tinyurl.com/6pjjxp.

5. Roger Penrose, *The Road to Reality. A Complete Guide to the Laws of the Universe* (New York: Vintage Books, 2004), and the references therein.
6. Franck Close, *Antimatter* (Oxford: Oxford University Press, 2010).
7. Philip Ball, "Did Einstein Discover $E = mc^2$?" *Physics World* (2011), August 23, http://physicsworld.com/cws/article/news/2011/aug/23/did-einstein-discover-e-equals-mc-squared.

Chapter 6
1. Douglas Arnold and Jonathan Rogness, "Möbius Transformation Revealed," last updated February 14, 2009, www.ima.umn.edu/~arnold/moebius/.
2. M. B. Green and J. H. Schwarz, *Phys. Lett.* 149B (1984): 117.
3. P. Candelas, G. T. Horowitz, A. Strominger, and E. Witten, *Nucl. Phys.* B258 (1985): 46.
4. Michael B. Green, John Schwarz, and Edward Witten, *Superstring Theory I: Introduction* & *Superstring theory II: Loop amplitudes anomalies and phenomenology* (Cambridge: Cambridge University Press, 1998).
5. E. Witten, "Reflections on the Fate of Space-Time," *Physics Today* (1996): P. 24, http://dx.doi.org/10.1063/1.881493.
6. A general mathematical idea of "point at infinity" is nicely given in *Unknown Quantity* by John Derbyshire (New York: Plume, 2006).

7. Clifford Johnson, *D-branes* (Cambridge: Cambridge University Press, 2003) (a reference title).

8. An application may further illustrate the concept of a self-contained system. In theoretical simulations of biomolecules—say, of a protein—a condition known as the periodic boundary condition is applied mainly to minimize artifacts that flare in in-silico analysis. Notwithstanding the designation "periodic boundary condition," the condition actually reflects the boundaryless state. The boundaryless state is applied employing the same mathematical functionality that is utilized in the construction of a Klein bottle or torus. The imitative cell that contains protein, water, and any other user-included ingredient lacks the actual boundary and exists in continuation by space-warp. The introduction of a boundaryless continuum eliminates artificial strain that the presence of a fixed barrier would cause, where the bulky process remains computationally efficient at the same time. In the periodic boundary condition, atoms do not leave the structural space, which imitates a self-contained fabrication, and yet goes on totally mobile. Atoms freely move fully under the influence of the set forces, and thus the observations refer only to the questions of the investigation.

9. Brian Greene, *String Theory on Calabi-Yau Manifolds*, February 23, 1997, http://arxiv.org/abs/hep-th/9702155.

Chapter 7

1. Jean van Heijenoort, ed., *From Frege to Gödel: A Source Book in Mathematical Logic, 1879–1931* (Cambridge, MA: Harvard University Press, 1967).

2. The School of Mathematics and Statistics, University of St. Andrews, Scotland, gives a concise account of set theory, including a summary of how Georg Cantor proposed the existence of the one-to-one correlation between the countable natural and rational numbers: http://www-history.mcs. st-andrews.ac.uk/HistTopics/Beginnings_of_set_theory. html.

Chapter 9

1. Ian Stewart, *Does God Play Dice? The New Mathematics of Chaos* (Oxford: Blackwell Publishing, 2002).
2. A representative title: Amit Goswami, *The Self-Aware Universe* (New York: Tarcher, 1995).
3. E. Witten, "Reflections on the Fate of Space-Time," *Physics Today* (1996): P 24, http://dx.doi.org/10.1063/1.881493.
4. E. Witten, "Duality, Space-Time and Quantum Mechanics," *Physics Today* (1997): P 28, http://dx.doi. org/10.1063/1.881616.
5. A general article on quantum histories (i.e. space-time streaming through all possible histories):
    M. Pössel, "The sum over all possibilities: The Path Integral Formulation of Quantum Theory," in *Einstein Online* 2 (2006): 1020, http://www.einstein-online.info/spotlights/ path_integrals
6. B. Greene, *String Theory on Calabi-Yau Manifolds*, February 23, 1997, http://arxiv.org/abs/hep-th/9702155.

Chapter 11

1. Popular articles on dark matter and dark energy:
    M. D. Lemonick, "Telescope to Hunt for Missing 96% of the Universe," *Time*, February 20, 2013.

M. D. Lemonick, "Can Tiny Galaxies Explain Dark Matter?" *Time*, September 17, 2013.

M. D. Lemonick, "Dark Energy: Was Einstein Right After All?" *Time*, February 28, 2011.

2. Richard Panek, *The 4% Universe: Dark Matter, Dark Energy, and the Race to Discover the Rest of the Reality* (New York: Mariner Books, 2011).

3. Mario Livio, *The Accelerating Universe: Infinite Expansion, the Cosmological Constant, and the Beauty of the Cosmos* (New York: John Wiley & Sons, Inc., 2000).

Chapter 12

1. A wormhole is a speculative feature of space-time that involves rupturing the fabric of reality. For a general idea of a wormhole, see the popular article *Wormholes as Time Machines* by Paul Halpern at http://www.pbs.org/wgbh/nova/blogs/physics/author/paul-halpern/.

2. The subject of the instances of space-time tears and ruptures in the framework of string theory is neatly detailed in the popular title *The Elegant Universe* by Brian Greene (New York: W. W. Norton and Company, Inc., 2003).

Chapter 13

1. "Conservation of Mass," *Wikipedia*, last modified May 1, 2015, http://en.wikipedia.org/wiki/Conservation_of_mass (and the references therein).

2. Some well-received texts on spirituality, and philosophy of the ultimate nature are as follows:

George I. Gurdjieff, *Life is Real Only Then, When 'I Am,' All and Everything* (London: Penguin Books, 1991).

George I. Gurdjieff, *Views from the Real World: Early Talks of G. I. Gurdjieff* (New York: Penguin Compass, 1984).

Osho, *The Mustard Seed*, (London: Element, 2004).

Osho, *The Way Beyond Anyway: Talks on Sarvasar Upanishad* (Pune: A Rebel Book, Tao Publishing Pvt., Ltd., 2001).

Osho, *Heartbeat of the Absolute: Discourses on the Ishavasya Upanishad* (London: Element, 1994).

Dennis Waite, *The Book Of One: The Spiritual Path of Advaita* (New York: O-Books, 2010).

Jiddu Krishnamurti, *Krishnamurti: Reflections on the Self* (New Delhi: BLUEJAY Books, 2004).

Sogyal Rinpoche, *The Tibetan Book of Living and Dying* (San Francisco: HarperSanFrancisco, 2012).

3. Jiddu Krishnamurti, *Reflections on the Self* (New Delhi: BLUELAY Books, 2004).

4. William Byers, *The Blind Spot: Science and the Crisis of Uncertainty* (Princeton: Princeton University Press, 2011) (The original article by Einstein appeared in the New York Times, November 9, 1930, pp 1-4).

5. Amir D. Aczel, *The Mystery of the Aleph. Mathematics, the Kabbalah, and the Search for Infinity* (New York: Washington Square Press, 2000).

Brief Notes: b. Chaos is a misnomer

1. For beauty, symmetry, and mathematics of fractals):

Benoit B. Mandelbrot, *The Fractal Geometry of Nature* (New York: WH Freeman and Company, 1982).

Ian Stewart, *Does God Play Dice? The New Mathematics of Chaos* (Oxford: Blackwell Publishing, 2002).

CREDITS

Image Credits

NASA images are credited alongside figures.

Calabi-Yau manifold (figure 6.2): A. Hanson, School of Informatics and Computing, Indiana University, www.cs.indiana.edu/~hanson.

Quotes

By Georg Cantor
First: As quoted in D. MacHale, *Comic Sections* (Dublin: Boole Press, 1993).

Second: Doctoral thesis (1867), Wikiquote, http://en.wikiquote.org/wiki/George_Cantor.

Third: As quoted in Rosemary Schmalz, *Out of the Mouths of Mathematicians: A Quotation Book for Philomaths* (MAA books, 1993).

By Kurt Gödel
As recorded in Hao Wang, *A Logical Journey. From Gödel to Philosophy* (Cambridge: MIT press, 1997). Courtesy of MIT Press.

By Albert Einstein
Chapter 5: Recalled by Leopold Infeld in Quest, 1941 [ed. 1980: p. 279]
Discontinuous Continua b., *Chaos is a misnomer*: From Einstein's letter to Max Born, 7 September 1944.

Courtesy of Albert Einstein Archives, Hebrew University of Jerusalem.

Chapter 5
Plato: *The Republic*, Book VII (360 B.C.).

Johannes Kepler: *Harmonices Mundi* (1618), *Wikiquote*, http://en.wikiquote.org/wiki/Johannes_Kepler.

David Hilbert: As quoted in *The World of Mathematics* by J. R. Newman, courtesy Dover Books.

Bertrand Russell: *Mysticism and Logic* (1918), courtesy Cornel University Library.

INDEX

www.ingramcontent.com/pod-product-compliance
Lightning Source LLC
Chambersburg PA
CBHW031819170526
45157CB00001B/119